728/1992

Android Design
Practical Approaches for Robot Builders

Martin Bradley Weinstein

HAYDEN BOOK COMPANY, INC.
Rochelle Park, New Jersey

Dedication

This book is dedicated to my beloved and devoted wife Judie, whose long hours of forbearance and tender hours of devotion helped make the work worth doing, and with whose help I will be developing humanoids long before androids.

Library of Congress Cataloging in Publication Data

Weinstein, Martin Bradley.
  Android design.

  Includes index.
  1. Androids—Design and construction. I. Title.
TJ211.W44   629.8'92   81-547
ISBN 0-8104-5192-1

Printed in the United States of America

4	5	6	7	8	9	PRINTING

83 84 85 86 87 88 89   YEAR

Preface

As of this writing, there are nearly a dozen individuals in America who have designed, built, and published plans for working robots of their own design. There are at least two companies now offering robot plans and kits as commercial products. The day has arrived when almost anyone can go to his hobby workshop and emerge with a home-built robot.

Still, we are reminded of the oft-quoted cartoon showing a hapless hobbyist who, having assembled a sailboat in his basement, finds that it won't fit through the door. Les Solomon of *Popular Electronics* told me of a robot with the same problem. Its designers had uncovered a novel, insectlike leg mechanism that permitted the critter ambulatory ease over any terrain, including stairways. The only problem was its limited ability to take sharp corners or fit through narrow doorways.

When you think about it, form often ignores function. Not even R2D2, the advanced droid of *Star Wars*, can negotiate a stairway.

The robots of fiction are often rather blithe in their ability to ignore design constraints. The robot of *Lost in Space*, for example, allegedly was powered by a small pack the size of a deck of cards. Even with batteries ten times as efficient as lithium cells (the best today's technology has to offer), such a power pack could operate a robot of that size at the speeds shown for less than 30 seconds.

Even the term *robot* is a bit suspect. In an early chapter, we will be looking at how robots, androids, and the ilk differ from each other. And while the admitted focus of this book is on *android* design, the builder of robots can glean a great deal of useful knowledge from these pages.

The design of an android is not the kind of project that fares well from a headlong plunge. There are many important decisions to be made at the outset, and a complex, interwoven electronic/mechanical/human engineering task to be accomplished.

You will find some plans and drawings in this volume, but what we offer here is neither a set of plans nor the final word in design. It is a thought-out attempt at organizing and recording one man's decisions on how to approach this design task.

You are encouraged to make your own judgments as to whether or not these decisions apply to the robot or android you build.

Branford, Conn.   *Martin Bradley Weinstein*

Contents

Introduction

I was surprised. I had expected people to ask me what an android is or how I planned to build one. Instead, they asked me what it would do.

What indeed?

My view of the problem was to first find a way to build a beast capable of doing *something*, then to worry about how it would do many somethings together—all long before I would worry about how to orchestrate the many somethings into the kind of answer my friends would want to hear.

Butler, babysitter, beanbag knitter, thief—how do you choose a career goal for a chestful of motors and microcomputers?

This is a book about design, not philosophy (not much, anyhow). Even just the task of carefully planning an android requires an exhaustive amount of interdisciplinary mind stretching.

We'll be examining what an android is, what we expect it to do, how that translates into design requirements, and how to bring a lot of loose pieces together into a whole.

We'll be looking at both usual and unusual hardware and software, mechanics and mechanisms. There are promising developments in some unusual corners that proffer opportunities for individuals to accomplish intriguingly sophisticated designs, and see them realized.

This is not so much a do-it-yourself book as it is a background on the tools, materials, and techniques necessary before designing your own android. And what you design need not look the least bit like anything I've managed to imagine.

But make no mistake, we're dealing with reality, not fiction. The designs we suggest are practical, accomplishable, affordable, and, with your help, open to a great deal of improvement.

1

1
Why Are You Reading This Book?

You can't actually want to build an android, can you? Do you even know what an android is? Do you have any idea of how complex a task you have ahead of you? The mechanics alone are boggling, the electronics nearly incomprehensible. And once you get the beastie built, what are you going to do with it? And even *that* assumes that you control the substantial sum of money you'll have to invest before you're through.

If you think this book is going to answer all these questions for you, forget it. At best, we'll be able to give you a slightly tighter grasp on the details and implications of each question, and maybe some options—strictly as a starting point—to help you make your own decisions.

Some professional wag once said that the solution to any problem lies in the defining of it, the answer to any question in the proper posing of it. So what we are going to try to do is to define some terms, set some goals, identify some tasks, examine current technology in a number of disciplines for approaches to these problems, and learn to draw lines between the desirable, the practical, the reasonable, and the doable.

Any ten people, working independently, who set out to build their own androids based on the guidelines in this book, are most likely to produce ten highly dissimilar mechanisms, each most right for his or her situation.

We are not examining facts in search of answers; rather, we are examining facts and goals in search of a discipline capable of yielding varying answers to varying problems.

Though android design requires no small involvement in a number of technologies, it is an engineering task that crosses traditional engineering boundaries. The distinction between engineering and technology is a fundamental one: while technology is a reflection of the tools, techniques, and equipment available to solve a problem, engineering is a reflection of the methodologies, disciplines, and procedures necessary to approach the problem. Past the Erector Set level, design becomes the work of the engineer, not the technician.

Fortunately, with some help from the manufacturers of the products of today's technology, so much good engineering information is available in such good form that you don't have to be a graduate engineer to design an android. If you read and understand *Popular Electronics* and *Radio-Electronics*, you probably have all the engineering savvy you'll need.

Once the design of your beastie has been accomplished, you'll want to move along to your first working prototype. You probably won't be able to. Chances are your first few stabs at a prototype will have major problems; require major troubleshooting; and uncover everything from fouled-up wiring to trying to put two things in the same place, to some stupid little thing that means "back to the drawing board" while it teaches you both patience and a valuable design lesson.

I know. I've been there. Many times. And that's as embarrassing a revelation as I will ever want to share with you.

The task of designing an android is very much like the task of deciding on what you want in a new car—but without the advantages of manufacturing, service, or stated models and options. The android you're going to dream up has never been reviewed by a magazine, no one just down the street owns one or can recommend one, and there are no warranties.

In the time required to move this manuscript from my brain to a book in your hands, anywhere from one to five years will have passed. In that time technology will have provided dozens of shortcuts.

But for all of the advances we can't predict, for all of the delays involved in getting new information to you, for all of the improvements that minds better than mine will discover, for all of this, you are still reading information that will never be obsolete. Because we're not giving you answers so much as we're showing you where to look for them and how to find them.

You still don't know why you're reading this book, though, do you? You just may have to read all the way through it, put it down and away for a while, then pick it up to read again before you can answer that question.

Other, more pressing questions come first. Do you really want to build an android? It doesn't matter if you don't—there's still plenty of good information here that can help you approach engineering, technology, perhaps even psychology from a slightly different perspective. For example, if you could choose only a dozen or a hundred or a thousand words, and those would be the only words you could speak or understand, which words would you choose? How would your choices change if there was some chance life might be at stake? How again if the operation of a car were involved? Or a plane? Should the recognized vocabulary be the same as the spoken vocabulary? How hard—or expensive—is it to add more words, or to change words? Must one voice be recognized, or many?

The problems of speech recognition and synthesis may never enter into the design of your android. On the other hand, they may enter into dozens of other aspects of your work or your life. Our discussions on the topic will be less than definitive but substantial enough to provide a broad background, identify the technology available, and show you where to look for more information.

So if you really do want to build an android, while we won't exactly be showing you how to *build* it, we will be showing you what's involved in the *design* of it—with examples, but without answers. And where we do slip and presume to offer an answer, feel free to disagree.

What if you don't really want to build an android? That's fine, too. Because you'll learn the basics of a lot of other things you may prefer to build, like electric vehicles, motor controls, alarm systems, prosthetic devices, movie props, collision-avoidance devices, and lots more.

In certain areas, some of my investigations so far have proven fruitless. It seems that technology is a least a decade away from permitting us a beastie who is anything but illiterate. Regardless of how resourceful he may be, it seems unlikely he'll be at all creative, despite the best efforts in the field of software art.

These problems, in turn, will greatly inhibit the eventual capabilities of even a very highly advanced android. Imagine an android bartender, unable to read the labels on the bottles: Anyone for a cognac martini or a creme de menthe/lemon/seven? The same predicament befalls an android chef, an android postman, or an android pharmacist—none of which we're likely to see in the near future, alas.

It would be immeasurably simpler to build the android equivalent to a dog or cat—but why would you want to? As a simple house pet, an android is much too expensive to be practical. If money were no object, nor the time necessary to pull everything together and accomplish the necessary miniaturization to accommodate a canine- or feline-size package, the project is not without its reward. An android pet could fetch the paper and your slippers, warn of newcomers or smoke or fire, perform a limited repertoire of chores and errands, and charm the bejabbers out of your guests. But it can't love. On the other hand, it won't have to be housebroken or walked or fed, merely refueled or recharged.

The point again is that you can now enjoy the option of being able to design an android to suit your particular circumstances. Whether you never get past the pencil-and-paper stage or go all the way into manufacturing the beasties, you'll be able to find your starting points here.

If you shop for books the way I do, you've already decided, on the basis of what you've read on the front and back covers and in the table of contents, that there will be something worthwhile for you here. And, in response to the question that is the title of this chapter, "Why Are You Reading This Book,?" you may have wished to pose a question of your own: Why would this be the first chapter of a book on android design?

Well, where would *you* start? We have a lot to cover, and any of a dozen starting places are serious contenders for the privilege; none of them, unfortunately, can give you as much of an overview as can your own curious intellect. And in our investigations, nothing is so valuable as a curious, incredulous, questioning intellect.

I encourage you to take nothing for granted.

2
Defining Terms

It's time to meet the several members of the genus and species we refer to variously as *robot, android, automaton,* and more. I've been through a mountain of reference material during the preparation of this book but have yet to find a single, definitive family of descriptions and definitions that isn't absolutely contradicted by some alternative source. So it is up to us to define our own terms. But we'll have to be careful, while maintaining consistency for ourselves, not to impose our definitions on the work of others.

As an intelligent first step to the task of arriving at some intelligent definitions, we should identify *which* are the key considerations. This, in turn, requires an empirical look at what our experience, our fiction, our plans, and our imagination tell us might be considered to fall under one of our definitions.

What then are the characteristics, in the broadest terms, that a mechanism must possess in order to be considered a robot, android, or whatever? What makes these machines different from a can opener or a computer—or aren't they different?

We have to draw a line somewhere, or we'll soon be calling a magnet a "robotoid paper clip tender." And, if we can draw a *few* lines, we can divide, identify, and define our species accordingly.

For example, while we wouldn't usually think of a computer as being a robot, we wouldn't hesitate to think of a robot as including or containing a computer—or attached to one. The difference is *not* one of size, because we can easily imagine a very large robot with a very small computer, or a very large computer operating a very small robot.

Instead, we find a division as old as literature: one is a thinker, the other a doer. In one we find the Apollonian nature of man; in the other, the Dionysiac. The first may consider information and make thoughtful decisions, but the second has the ability to take action, to manipulate itself and the things it comes in contact with.

So, as a first part of our definition, we have been able to identify a *manipulative imperative*: an ability to manipulate objects external to itself is a requirement for all classes of equipment we want to consider. Still,

7

while this may exclude pocket calculators from our definitions, it doesn't exclude even the lowly automatic record changer. Should it?

It may seem obvious to moviegoers and science fiction buffs that a record changer doesn't move around and robots (etc.) do, but in this case facts put a lie to fiction. The images conjured up by the droids of *Star Wars* quickly give way to General Motors of Lordstown, Ohio, the site of a fully automatized automobile assembly plant. This factory is the robotic analogy of Eli Whitney's and Henry Ford's accomplish- ments with replaceable, interchangeable parts, and production lines. Here and throughout manufacturing, industrial *robots* (so-called by their manufacturers) are working daily, working on fixed bases as they reach, swivel, screw, and spray.

Mobility, then, provides one of those dividing lines-between- species we hoped to find. We will have to look elsewhere for reasons to exclude that record changer—if, indeed, there are any.

Is there something about a record changer that's especially con- fusing? Perhaps we need to define what it is and isn't before we can differentiate it from our other mechanisms. As an exercise before you read on, you may want to try describing the actions of a record changer yourself. Treat it as a totally new device you want to explain to someone who has never seen one. Let's see . . .

It accepts a variable quantity of phonograph records, which are the raw materials external to itself it must manipulate. It may be able to accept several records of various diameters, or it may require pre- setting to one size by its operator. In every case, the operator is re- sponsible for loading the records, selecting the sequence in which they are played, selecting the speed at which they are played, and initializing the automatic sequence. The changer then senses whether or not there is a record loaded. If there is, then only one record is dropped onto the turntable; if there is not a next record, the changer goes through the sequence of resetting itself to a rest position and turning itself off.

Before a record drops to the turntable, the turntable motor is started and the tone arm lifts up, out of its cradle, and is held to one side, out of the path of the dropping record. After a brief delay, which allows the record time to drop and come to speed, the tone arm swings inward along an arc until it is poised just above the first band of the record, which is a molded ramp that guides the stylus into the record groove spiral. At the completion of its journey through the spiral, a position sensor begins an end-of-record sequence.

Now, the tone arm is first lifted from the record, then swings to a "ready" position fully to one side, usually just above its rest position and cradle. Again, the changer checks to see whether or not a next record is loaded. If so, it repeats the play cycle; if not, it begins a power- down sequence. The tone arm drops into its cradle, the turntable motor is turned off, and the drive mechanism is disengaged.

So here we have a mechanism that, once supplied with materials and started, automatically accomplishes a set task, makes decisions based on observations of the external objects it manipulates, and adapts its actions accordingly. Its response to any given situation will be the same every time. It is consummately predictable. It does not learn from experience. And while controls may be varied to produce variations in its responses, there is no facility for reprogramming it to permit a new and different set of predictable responses.

Here, at last, is a difference between our record changer and the large industrial robots: *reprogrammability*. The programming that is in control of the set of available manipulative options in the latter case can be updated at any time. Reprogrammability is one of the major keys we've been looking for.

Our previous description of the workings of a record changer contains two other key thoughts in one phrase: ". . . makes decisions based on observations of the external objects it manipulates, and adapts its actions accordingly." This is the source of our hesitation about excluding a record changer from all the classes of mechanisms we want to include as robots, androids, automata, and so forth. The ability to make observations, the ability to make decisions based on these observations, and the ability to adapt manipulative responses according to these observations and decisions are all significant to our eventual definitions. These also are important considerations in the classical definitions of *intelligence*.

Once we assert that mechanisms that manipulate objects external to themselves in an entirely random fashion are of no interest to us (you may wish to spend some time considering the validity of this statement, but it is one we must state if our definitions are to be reasonable), we must recognize that there is some source of intelligence in control.

The nature and location of this intelligence provide another class of considerations in our definitions. Is the mechanism under the constant control of an external intelligence? Of a self-contained intelligence? Or is it only occasionally instructed, updated, or reprogrammed? Perhaps there are other divisions as well. The nature of this *control* is a fundamental key to our definitions; interestingly, the magnitude of intellect of the controlling intelligence is not.

We've already noted an ability to *adapt* is a significant capability, which leads us to consider the capability for *learning*. Of course, since we've also specified *reprogrammability* as a key, we recognize that a new behavioral perspective can always be force-fed, but we're much more concerned here with the mechanism's own ability to use its experience as a factor in making decisions.

For example, if the mechanism is designed to follow certain guidelines when faced with decisions, but to weigh these rules of thumb against the successes and failures of its own experience in similar situations, the original guidelines may eventually be totally supplanted by the lessons of the mechanism's experience. Unless forbidden or constrained from doing so, the mechanism may even discover, create, or invent new answers to any given decision-making problem. In education, the terms *heuretic* and *heuristic* are applied to this kind of learning. More to the point, *heuristic* programming is a rapidly developing aspect of software craftsmanship. So, if *learning* is possible—and we know that it is—we must consider it in our definitions.

One final consideration before we can at last come to terms with our definitions is whether or not the mechanism is designed to closely resemble humans. This is not as frivolous as it may sound.

We have designed the constructs of the world around us to fit our own forms, and they will not tolerate a great deal of variation. Robbie, the affable mechanoid of the sci-fi film *Forbidden Planet*, could never fit in the driver's seat of your car, pilot a jet fighter, reach deep in the back of the bottom shelf of your pantry cupboard for a can of soup, or follow you through a revolving door.

If the Air Force could use nonhuman test pilots, lives might be saved. But planes are designed for beings with the shape, reach, and dexterity of a human. So *resemblance to humans* becomes significant.

Let's see if we can pull the pieces together now and juggle them into some understandable order. In so doing, I'll present my definitions as to which beastie goes by which moniker:

Automaton A fixed or mobile mechanism capable of manipulating objects external to itself under the *constant control of a programming* routine *previously supplied* by an external intelligence, capable of being reprogrammed, and capable of adapting its actions according to decisions governed by its programming and by observations of the objects it manipulates.

Robot A fixed or mobile mechanism capable of manipulating objects external to itself under the *constant control of* a human, a computer or other *external intelligence*. It is a remote mechanical manifestation of the commands of its controller, capable of adapting its actions according to decisions governed by its controlling commands and by observations of the objects it manipulates.

Android A mobile mechanism capable of manipulating objects external to itself under the *constant control of its own resident intelligence*, operating within guidelines initially established and occasionally updated through reprogramming by a human, a computer, or other external intelligence. It is capable of adapting its actions according to decisions governed by its programming and by observations of the objects it manipulates. It is capable of limited self-direction and initiative when not involved in program-mandated tasks.

Cyborg   (from "cybernetic organism") A fully independent android capable of heuristic learning and self-updating of its own resident programming intelligence. An advanced mechanism, bordering on *being*, only marginally accomplishable with current technology.

Mandroid   An android or cyborg in the shape, size, and likeness of a human. Size, power, and dexterity limitations of current technology make this mechanism only marginally accomplishable.

Bionic   Not relevant to our discussions, but included here in the interest of clarity. This term is used in reference to technology prosthesis—electronics and mechanisms used to replace or improve the operation of senses, organs and body parts, and cosmetically identical to their biological counterparts.

Mechanoid   Any mobile mechanism capable of manipulating objects external to itself; a subclass of automaton, robot, android, or cyborg.

Robotoid   Referring to any mobile mechanism capable of manipulating objects external to itself. Compared to mechanoid, robotoid implies more sophistication, more intelligence, more electronics, a less mechanical construction. Neither term will be used widely here, as neither is very specific.

Our topic at hand, of course, is *android design*. And now that we have a handle on what an android is and isn't, we can get on with the task.

3
The End

This is a good time to give you a peek at what lies at the end of the design task—a finished, functional android. Mine and yours are both years short of the following description, but it is an entirely *feasible* description, with current technology. Which is to say that if we had infinite money and assistance, we could build this android with off-the-shelf components.

Imagine a mechanical contraption about 4-1/2 feet tall from the base of his tracks to the top of his eye stalks. The twin track base is under 3 feet long by under 2 feet wide by about a foot high.

A cylindrical trunk extends upward from the center of the track chassis. It's capable of leaning forward or backward as much as 45°. An automatic subsystem keeps it upright; this can be overridden if the android purposely decides to lean.

Two arms extend from shoulders at the top of the cylinder. The sizes and locations of the motor and drive boxes gives the arms somewhat the look of a mechanical Popeye the Sailor Man.

In detail, the shoulder box is connected to the arm lift box; the arm lift box is connected to the arm rotate box; the arm rotate box is connected to the forearm lift box; the forearm lift box is connected to the forearm rotate box; the forearm rotate box is connected to the wrist lift box; and the wrist lift box is connected to the hand.

The hand is over a foot long (punsters, control thyselves!) from wrist to fingertips. There are three fingers opposed to two thumbs on each hand. Unlike human hands, the base of each digit is a hinge, rather than rotary joint. The three-opposite-two configuration allows a firm grasp on such things as hammers, brooms, and glasses. Sensors in the palm of the hand help the android guide its hands into the best position before closing its grasp.

The fingers themselves are marvels of ingenuity. The knuckles and bones are links of lightweight chain. Thin wires driven by motorized pulleys perform the flexor/tensor (also called flexor/extensor) muscular function. Squeeze bulbs in the fingertips connect through flexible tubing to pressure transducers to provide a portion of the android's sense of touch; this is augmented by switches between the "knuckles" that sense grasp pressure and by feedback from the pulley motors that indicate motor loading. Thermal sensors in the fingertips complete the android's sense of touch.

The two thumbs are driven together. The two outer fingers also are driven together. The middle finger is driven separately to allow it to perform such tasks as button pushing, dialing and ring pulling (as on window shades and talking dolls—no telling when the beastie will be called upon to babysit). The fingers are booted with oversized rubber gloves to provide a more natural feel, a frictional surface, and some degree of waterproofing.

The arms and hands are not only completely ambidextrous, they also are fully capable of working equally well on either side of the trunk. The trunk is capable of rotating through a 540° arc (one and a half revolutions) permitting 90° overtravel at both ends of a one-full-revolution turn. This allows full flexibility while easing the task of providing connections from the base through to the head. The head, too, is mounted on a turntable permitted and limited to 540° of travel.

The most prominent feature of the head is the twin camera mechanism mounted on it, looking somewhat froglike. Two identical cameras, each about the size of three packs of cigarettes in a stack, sit on tiny turntable bearings atop a rectangular box; the rectangular box is mounted on a U-shaped bracket and tilts up and down.

The twin cameras are driven by a linear actuator to give the android parallax vision. This, in concert with information from the tilt mechanism, permits it to accurately locate points in space. This is augmented with range information from a SODAR installed on the front face of the tilt box. A headlight on the same face helps the android get around in the dark—hopefully just long enough to turn on a light in the room.

Each camera "eye" performs a different set of tasks. Between the two (and associated processors, memory, and so on), they're capable of recognizing objects, locating obstacles and free paths, and coordinating hand and arm movements. Processed information also is inputted to a subsystem that provides, stores, recalls, and seeks matches against "maps" of the room the android is in, as well as some number of rooms the android is customarily in.

The cameras are arranged to look in parallax at varying angles to cover a range from a few inches to infinity. The tilt mechanism can rotate from a few degrees upward to 90° straight down.

A speech box behind the vision components houses a speech synthesizer, amplifer, and speaker. A vocabulary of just a few hundred words is highly flexible and adequate enough for a household android. There also are provisions for teaching the android how to pronounce new words.

Hearing boxes are located on each lateral side of the speech boxes; these house quasidirectional microphones and preamplifiers. The hearing boxes perform two functions.

First, each microphone drives a different kind of speech recognition circuit. Outputs of the two speech recognizers are compared against each other for good real-time comprehension; the recognition vocabulary is, of course, limited, but quite flexible with as many as several hundred words included.

Second, phase information from the two inputs is compared to try and locate the source in azimuth; if the other sensory systems are not otherwise encumbered, they may try to confirm the deductions of this subsystem in an effort to "train" it through heuristic self-improvement.

The android literally bristles with object detectors. A variety of approaches to object detection is used to assure that objects transparent or invisible to one are detected by another, and objects detected by one can be verified by another.

SODAR is one approach. Ultrasonics are used, with the time necessary for the front edge of a burst to reflect from an object and back to the transducer yielding range information. Five SODARs are planned. Four of them will be placed 90° apart near the top of the torso cylinder, the fifth on the vision tilt box. These five are activated sequentially with appropriate delays to assure that they do not interfere with each other. Also, since there may be multiple reflections, only the first arriving burst edge is used by any SODAR. The four torso SODARs are used primarily to locate walls and large objects for both the collision-avoidance and mapping subsystems. The fifth SODAR is used for range information for the visual system; also, because the head can rotate independently of the torso, this fifth SODAR performs a number of confirmation and backup functions.

First, recognizing that the torso rotates independently of the drive chassis, there may at any time be no forward-facing SODAR; the vision SODAR can be rotated with the head to perform this function. Second, this fifth SODAR can rotate to find the minimum range to an obstacle—the perpendicular distance—to reduce or eliminate ranging error as the android seeks to locate its position relative to walls as part of its mapping function. Optionally, the rotational position of the head can be used to correct the rotational position of the torso for complete measurement of room dimensions and position within those dimensions.

SODAR (it stands for sonic detection and ranging) is the equivalent of SONAR, except air rather than water is the carrier medium. Ultrasonic frequencies are used because they dissipate less quickly, bend less, can detect smaller targets, and need smaller transducers than lower frequencies. The frequencies actually should be high enough to be out of the range of hearing of dogs and other pets, and selected to not interfere with ultrasonic remote control frequencies that may be used for other purposes within the android's intended home. As a side benefit, chances are good the frequencies you use will repel flies and mosquitoes. And note that SODAR will not be fooled by large windows, as a strictly optical system might be.

New developments in SODAR by camera companies may prompt the use of sophisticated but inexpensive multiple-frequency "chirp" signals; the techniques used are similar, regardless.

A second class of object detectors is essentially photoelectric. A narrow beam of light, pulsed at a preset frequency, illuminates objects in its path; a sensitive photodetector is focused along the same narrow path. The output of the photodetector is compared to the frequency modulating the light source in a phase-locked loop tone decoder. A minimal delay is built into the detection circuit to help minimize false readings.

Where permissible and desirable, low-power diode lasers can be used as the light source; elsewhere, focused high brightness LEDs are more desirable.

Groups of 4 sensors are located at the top of each of the four corners of the track chassis, at 30° increments from straight out the end to straight out the side. A second group of 16 sensors rings the top of the torso at 22.5° increments. A third ring of 16 is near the base of the torso but clears the top of the track assembly. A line of 4 sensors near the floor at each end of the carriage chassis looks for small objects on or near the floor; a second line above the first looks ahead of the first, making a total of 16 sensors in the last group.

The frequencies chosen for each sensor can be selected so that no one can mistakenly read its neighbor; sequential switching also is possible. Modulation of the transmitted and reflected light helps assure that external light sources can't fool the sensors. As a failsafe, a system of bumpers incorporating ribbon switches and microswitches sense actual contact with objects.

Power for the android comes from an on-board battery. The battery's charge level is constantly monitored; when low, the "hungry" condition is sensed and a feeding routine begins.

The first goal is to locate a suitable electric outlet. The android's map of the room is consulted. (When first entering a room, the android's programming is designed to assign mapping a high priority. This includes locations of walls, obstacles, doors, and outlets.) If no outlet location is registered, a search for one begins immediately; if an outlet location is known, it is sought out.

A highly sensitive amplifier with peaked power line frequency response and a small antenna with quasidirectionality at each of the four top corners of the carriage chassis seeks out the hum fields that house wiring radiates. Comparison of the hum level at each corner helps locate power lines and power cords. The visual system confirms outlet locations.

The battery charger and a power cord on a reel are contained within the android base. When the outlet has been located (where possible, one not being used; otherwise, the android will use the level of remaining charge as a guide in deciding whether to keep searching or to pull a plug—although certain plugs, like alarms and medical equipment, can be marked and the android forbidden from pulling them) the android will plug in and enter a "feeding" condition.

The android continues feeding until the battery has reached full charge, although nonmotor functions remain fully active and the android can decide that a threat is present; under this condition, the android will avoid or prevent the danger, as appropriate, before returning to feed.

When feeding has been completed, the android unplugs, stores the power cord on the cord reel, plugs back in any plugs that have been removed from the outlet, and goes back to work. The feeding mode also can be entered as the result of a command or decision to enter an inactive mode.

If no power source has been located and the charge remaining reaches a critical level, the android enters an energy conservation mode, cutting all motor functions and calling for help. This cannot happen on stairways or places where power-down could cause a hazard.

The brains of the android are located all over its body, with the primary overseer processor located inside its cylindrical torso. Dozens of microprocessors are used in an organized synaptic system.

Synaptic organization of the processors means that levels of intelligence and control are nested in layers. Each processor is assigned a specific task, beginning with the lowest intelligence level processors that condition sensory inputs. The lowest levels of processors communicate with, report to, and are controlled by only one processor at the next level up; up the chain, while a processor may (and probably does) communicate with and control several lower-level processors, it too communicates with, reports to, and is controlled by only one processor at the next level up—and so on, up to the overseer processor.

This strict chain of command both avoids conflicting instructions to a single processor and eases the development task. It also reduces troubleshooting time and effort in the event of a problem and makes failsafes easier to implement.

This approach also modularizes the design task. As a new approach to each task emerges with emergent technology, it also permits easier updating of the android. Also, by requiring *stated communication formats* (microcomputerists may prefer the term *defined bus structure*), it encourages commercial development of the various subsystem modules.

Now that you've had a glimpse at what lies at the end of the design task, let's get back to getting it started.

4
Philosophical Considerations

It is said that our fiction mirrors our future. Bunk. Our fiction mirrors so many versions of our future that any number of them could be part right, yet all wrong.

In my role of researcher, preparing the materials for this book, I felt duty bound to read *I, Robot* by Isaac Asimov, the science fiction classic that first introduced his now-famous "Three Laws of Robotics" to the world. His world. A world of plot and resolution. A world of delightful storytelling, captivating stories, excellent writing, and pleasant reading. Which is not in any measure intended to demean the very talented author's abilities in science as well as the arts.

Unfortunately, we are generations away from being able to produce a mechanism anywhere near as sophisticated as his fiction offers. So it is with apologies and a bow to Asimov that we begin our task of stating ethical restraints for the design of an android in developing our own three laws of "androidics."

As in defining terms, our best beginning is to thoroughly recognize how our beastie reacts with his contextual world. Rather than prime directives or chains of command, let's start by designing our own "environmental impact study." This, in turn, requires that we anticipate somewhat the famous laws of Mr. Murphy and examine as bad a worst-case condition as we can imagine.

The absolutely worst case is one in which all controls get hung up in a maximum speed, maximum power, minimum control mode. Contumaciously running amok, what damage might be caused? To what? To whom? Should we set some priorities, so that when options exist, we might choose one form of destruction to be "better" than another?

Obviously, you don't want to build anything that's going to kill, hurt, or damage you, your friends, your family, or pets. The lessons of history also have taught us that it is unwise to allow anything of this sort to happen to other living beings, including animals and—once we extend the argument—plants.

Ahh, but is that a universal case? Is it better to fight off or attempt to subdue an attacking dog, for example, with an expendable mechanical contrivance, or to attempt the hazard yourself? The answer to that lies in our worst-case no-control condition. While you may indeed wish to build an android to protect yourself against lions and tigers and bears (oh my!), it will be at your command, not by an accident of your creation gone wild. The question of judgment—of knowing that unless action is taken, you are in danger, and only through the subjugation of a living being can your safety be assured—is one we must wait a bit to address.

Should we also be concerned with nonliving things? Even beyond the *manipulative imperative,* which mandates that an android, in order to be considered one, must interact with objects around it, our mechanism is bound by physical and practical considerations to coexist with us in *our* world. Which means doors, walls, carpets, floors, and furniture. Shoes and clothing. Windows, dishes, glassware, and that bottle of twelve-year-old firewater you've been saving for a special occasion. Unless you can relegate your android to some remote, undamageable corner of your life, the chance that it will damage property always exists, even with the best of efforts to prevent it.

While we're at it, let's consider some of the less tangible but nonetheless equally real aspects of our environment, like the air we breathe, the temperature around us, and the relatively hygienic way we prepare our food. Like our first consideration, we cannot let these become the channels by which our mechanism can do harm—even when totally out of control.

So, again with apologies to Asimov, let us now propose our first three laws (*rules,* more appropriately) for android design:

1. First, our creations should not be destructive to any part of their environment, including living cohabitants as well as walls and furniture, the breathableness of the air, radiation levels, anything.
2. Second, our creations should not be destructive to themselves. So we need to include adequate hardware and software protection to assure self-preservation, except where this violates the first "law."
3. Third, we must design in an instinct for survival, which translates to self-continuance of operation. The most immediate manifestation of this trait (by way of example) will be a mechanism to assure that low batteries will be recharged before failure.

The term *mechanism,* here and throughout the text, will be used to identify any and every means—hardware, software, human cooperation, the assistance of another android, whatever—by which an end or a design goal may be achieved.

Our first "law" neatly summarizes the discussion so far, but again with no reference to *judgment* or judgmental priorities. The second "law" is a natural extension of the first, which assures that we don't build beasties that do themselves in. And the third "law" neatly extends the second to cover sins of omission, as well as commission.

Unfortunately, these three "laws" are of little help in anything but an overview of the nature of the hazards of property damage and bodily injury of which an uncontrolled android is capable. A more realistic exercise would be to list a longer roster of more precisely defined hazards, as exhaustive a list as possible. And it helps to begin with the item most valuable to you: *your life*. The progression from here is readily apparent: *any human life, any mammal life*. But at this point, priorities are a bit more moot. You may decide to modify the order of the following list.

Injury to a human. Death to a mammal. Danger of fire. Injury to a mammal. Danger of noxious fumes. Damage to property. Damage to android. Creation of access to dangerous, potentially dangerous, or hazardous materials or circumstances.

That's the theory. In practice, we're talking about things like battery acid, battery gases, overheated parts, spilled or sprayed lubricants, vaporized hydrocarbons, exposed radioactive materials (even smoke detectors contain them), and bursting light bulbs.

Since we can't give an android conscience and religion, per se, we must design the Golden Rule in as best we can. Or more specifically, we must make certain that when it comes to *doing* unto others, the android *doesn't*.

When man evolved from the lower orders, and from an individual into a social animal, we had to trade certain personal rights for our new social rights. This has been a bone of contention ever since, with Thomas Paine and Edmund Burke stating opposite sides of the case quite elegantly as recently as the American Revolution.

Our android must evolve immediately (which makes the word "evolve" inappropriate) as a social animal, or at least as a societal beastie. We are the society into which an android must be integrated. His prime directive must be: don't hurt.

But we are making a fundamental error. Acts of conscience by an android are out of the question; he has none. All of the acts of conscience are on our parts, in the design stage, with our creation at best a passive follower and interpreter of our guidelines.

For example, there is not a great deal of difference between letting an android loose on his own and releasing the parking brake of a truck on a hill; whether or not anything bad happens depends more on the failsafes designed in than on the controls.

In the early days of steam power, railroads quickly learned to install what is called a "dead man." This is a control designed so that any time the engineer takes his weight off of it—and the design usually combines a lateral or rotary motion, to make sure that maintaining pressure on it is no accident—power is automatically cut and the brakes applied. This particular failsafe was incorporated to keep a train from chugging ahead at full speed with a dead engineer—or no engineer— at the stick.

You could achieve the same result through the combined efforts of hardware and software, or with hardware alone. You cannot trust software alone, since a software hangup could prevent software failsafes from working. One simple scheme would take advantage of the versatile 555 timer, used as a retriggered monostable multivibrator in a missing pulse detector configuration, a standard circuit published by most of the manufacturers of the 555 in the applications section of the data sheet. The 555 would require a new triggering signal, under software command, before the monostable RC timing circuit times out. This could be set to something like twice a second, a long enough interval to prevent the updating task from becoming too tedious a burden on the software, yet a short enough interval that the android could do no real damage before the failsafe takes over.

The failsafe itself will require no small amount of deliberation. You wouldn't use a fuse on a stick of dynamite as a failsafe device, but the same effect might be realized if your failsafe results in a short across the battery, for example. Even a short near the battery, with its resultant high heat, could be dangerous.

One option might be to connect the battery through a high current switch that is mechanically manipulated by a solenoid. It would take a concerted effort to mechanically engage the switch, thereby energizing the solenoid (alternatively, you could "jump start" the solenoid) when engaging the android, but any failsafe could de-energize the solenoid, dropping the switch out and requiring your intervention. Parallel switches also might apply the brakes, short the motor windings, sound separately powered alarms, or radio ahead to the junk yard with a reservation.

Other failsafes might take less drastic action. Collision avoidance, for example, can be accomplished through an array of various contacting and noncontacting sensors, suitable steering logic, and overrides in the hardware or software controlling motor speed and direction, braking, and so forth. The efforts of the automotive industry over the last several decades have contributed a great deal to the subject, and references are readily accessible.

Another suggestion borrowed from the automotive industry— especially in the years since the emergence of Nader's raiders—is the use of materials that *give* on impact; foam rubber is a lot easier on the knees than steel. Elastic materials and cushioning can contribute a great deal to the "neighborliness" of your android.

These suggestions are brought up here, remember, from the perspective of a developing *philosophy* of design. Simply stated, in all good conscience, our philosophy toward design *must* be that an android is intended to aid us, not to harm us, others, pets, or damage property. Further, as a contributing member of society, our android must be designed to continue contributing with as little attention as we dare give it.

You, as designer, will develop your own philosophy, possibly even to the point of ignoring these few bits of advice entirely—which hardly seems likely, but then, neither did the prospect of being able to actually build an android until just recently. Whether you agree with it or not, you must concede that the first point in our design process is to identify and state the assumptions that seem obvious to us, but still denote decisions.

Once your assumptions are in black and white, they'll need examination, restatement, embellishment, and thought. Assumptions beget ramifications. A particularly sticky example is our early assumption that we and the android will be living or working together.

When's the last time you tried driving through your kitchen?

5
Obstacles in the Human Environment

Getting around is something you haven't given too much thought to since infancy, in all likelihood. Walls and stairs and halls and doorways are not likely foremost in your thoughts, even when you're in the midst of them. That's because humans have a tremendous advantage—our environments were designed to accommodate us.

It's up to us, though, to design our androids to accommodate our environments, to exist in them without damaging them, to maneuver through them as naturally as possible. To fully appreciate the complexity of this task, let's embark on a little mental scenario for a fast experimental demonstration. Oh, yes . . . please *do* watch your step!

Since, from one point of view, our android will be a vehicle, let's begin by substituting a more familiar vehicle. (The android, after all, is rather nebulously defined.) A radio-controlled toy car will fill the bill nicely.

Now imagine that you are in control of one of these small cars as you take a walk around your home. Negotiating the clear paths is simple enough. But where are the obstacles? While we're at it, let's keep watch for ways in which we might use electronics to sense obstacles— or to sense clear paths, an equally valid approach. What kinds of things prove dangerous? Where might we lose control, face inextricable geometries, or find our drive system inadequate?

Troubles appear from the start, including troubles in starting. Hard, level floors can be slippery. Fast or high-torque starts and turns can leave marks. Rugs can snag or bog wheels. Edges and door sills can mean impasses or unplanned changes in direction. Uneven floors can do the same, as can exaggerated grooves between tiles in tile floors.

All of these problems can occur before we leave the room. Are bigger wheels, a bigger wheelbase, or larger cars an answer? Or will they solve some problems at the expense of others?

Also, flooring isn't the only obstacle that appears on a floor. Electrical power cords, speaker cords, and so on crisscross the fringes of our rooms. Pole lamps have bases, furniture has legs, and there's no telling what newspapers, magazines, toys, shoes, socks, or other paraphernalia may litter a given living space.

23

Door jambs and door stops can jut out from walls and doors. Doors themselves may be open and in the path of our imaginary car. Clock weight pendulums and chains may dangle down into its path without actually touching the floor. Not to mention the awesome cliffs of a stairway.

In addition to where objects actually *are,* we must consider where they *might be* next. (This concept of a *probability shell* will receive closer attention in our discussions of collision-avoidance mapping.) A table, for example, is highly unlikely to be someplace at one instant and someplace else at the next (although there is *some* chance of it, as for example when we move it). A house cat, on the other hand, is highly *likely* to be in a new place at a new instant (though there is a definite chance that it won't be).

So far in this imaginary journey, we have been the eyes and senses of the vehicles in our control. Soon, we will be forced to reconsider the journey with blindfolds on. But first, we have to consider the problems of size and drive.

Obviously, we can't make an android taller than our ceilings, wider than our walls, or heavier than our floors can support. These dimensional limitations are very much obstacles to the designer, albeit obstacles we're more likely to *design for* than to overcome.

Also, since the bigger we become the more obstacles we encounter (including the need for more powerful components and larger power sources), it is in our best interests to limit the scope of sizes we wish to consider. A careful consideration of the obstacles we are likely to encounter is one very effective, very sensible way of determining what maximums we should set for these dimensions.

As a short cut, you might conclude that anywhere we fit, an android the same size as a human will fit as well. In practice, though, there are good reasons to consider sizes bigger than our own—the difficulty of walking, for one, which results in the need for a drive larger than feet and legs.

Instead, let's try to identify those obstacles that specifically limit specific dimensions.

Height is affected by ceilings, of course, but doorways are more significant. In specific rooms, lighting fixtures also may prove a limitation; with lights, at least, it's possible to make a decision to treat them as limiting factors or simply as anticipated obstacles.

Width will be limited by doorways (don't forget to take into account the portion of a doorway occupied by an open door on its hinges), stairways, hallways, encroaching shelves and furniture, and so forth.

Length is going to be limited by the available turning radii in halls, doorways, stairway landings, and so forth.

Weight (physical purists may prefer to consider *mass*) is constrained first by the ability of supporting structures to bear weight loads, and second by the ability and necessity of the motive drive system to start and stop quickly.

After examining the specific environment in which you intend your android to operate, make note of the height, width, length, and maximum weight it will tolerate. These roughly hewn dimensions are only a starting point, but they are a good and practical starting point.

One remarkable difference between our android and a vehicle of any size lies in the manipulative imperative; while the primary goal of the car in our imaginary simulation is *obstacle avoidance*, the android is more likely to approach the task as *obstacle recognition, collision avoidance*, and *obstacle manipulation*. To illustrate the difference, consider a door. If closed, the door represents an impasse to a toy car intent on crossing the doorway; an android, however, has the option of opening the door.

These considerations of obstacle manipulation and dimensional limitations are important as we embark upon the second phase of our imaginary excursion, this time in an unfamiliar room that we cannot see.

To ease the task somewhat, we can consider that our model moves at a fixed rate of speed and is capable of accelerating and decelerating instantaneously. Also, we will assume that we have perfect control over its direction and control.

Our goal is to try to identify ways in which electronics can be used for obstacle recognition and what their shortcomings may be. A more complete, practical discussion of specific schemes will be undertaken later in our review of electronics for collision avoidance.

For the sake of this discussion, we will assume that our sensor is a "feeler"—a stiff wire, for example, connected to a switch—that lights a light on our control console whenever it touches something. And finally, we will assume that the vehicle we are controlling is a vertical cylinder, as wide and as high as a human. This cylinder also offers us a unique skin that signals us with a buzz at our console if it touches anything.

One approach to our obstacle-avoidance problem might be to purposely *not* avoid them. We could continuously drive the cylinder into the walls of the room to define a perimeter. Since it moves only at a known, fixed speed, a stopwatch and some graph paper could quickly yield a to-scale chart of the walls. Then, using whatever scheme we wish to devise, we can cover the rest of the room. The result is a complete map of all the obstacles in the room. Once that map is compiled, we can always steer around the obstacles from then on. Of course, a new map would have to be compiled for each room we enter. And an open door might, for a time, confuse us into believing that multiple rooms are actually one; granted, this is inconsequential since we eventually would determine a complete map in any case.

In real life, unfortunately, this "bull in a China shop" approach is rife with potential for destruction. Rather than crashing headlong into obstacles, we will be forced to make use of our feeler—or several feelers. The question becomes how few we can get away with while covering nearly all potential obstacles.

This is where we can take advantage of some of the lessons we learned on our first journey. If we first identify those specific obstacles we can anticipate encountering and design a feeler array capable of warning of them, we can then throw some random elements into the room and see how well our array copes with them.

Most of the time we are in forward motion. (In practice, this greatly eases requirements on drive systems, sensor systems, power sources, drive control systems, and elsewhere. For our purposes, however, we can consider this simply an assumption to simplify our model—also, "forward" can be considered an arbitrary but defined designation relative to the geometry of the android.) So we will certainly need some sensing in the forward direction—but how many "feelers" and at what height or heights above floor or ground?

Let's consider the obstacles we know we are likely to encounter: walls, furniture, electrical cords, doors, and so forth. One class of obstacles is small and likely to be in our path—these require feelers very close to the floor. Another class includes objects like furniture, with very little or no presence at floor level but a great deal at 2 to 3 feet up—one cleverly designed or cleverly placed feeler covering this level should catch most of them. A third class includes objects like transoms and chandeliers that could knock an android's block off, or be damaged by it—a top-mounted feeler will be necessary to sense these.

Can any of these levels be eliminated? One way to tell would be to look for traps—obstacles that might confuse our sensor array.

If we try to get by without our floor feeler, shoes without feet in them, wastebaskets, and toys are some of our traps. Another more significant trap is an open window, which would "read" the same as an open door. Without the middle feeler, even a chair might prove to be a trap; there's a good chance our floor feeler would miss a chair's spindly legs. Desks and tables are another trap, because their legs are widely separated (compared to the maximum width we've set for the android) and out of feeler range. The traps for the top feeler are those very obstacles we've included it to sense—chandeliers, transoms, and so forth.

Where else might this "trap" analysis lead us? Or, put another way, are there still traps for the three sensing feelers we've started with, obstacles they're likely to miss? And where else should we put additional feelers to catch them?

One characteristic of our feeler array so far that may prove significant is that it is one-dimensional, vertical only, while the android is three-dimensional (four-dimensional, actually, because we must eventually include *time* in our collision-avoidance parameters, through the incorporation of both probability shells and trajectory calculations). Since the android can move in at least two dimensions, spatially, and occupies three, we can look at the need for at least a two-dimensional feeler array.

Let's go back to our mental exercise, the "blindfolded" remote-control excursion through the house, looking for traps we might encounter with a single line of feelers. In fact, we can even start in an empty house, with no furniture.

We reach one potential trap with the first doorway. If the door is fully open or fully closed, no problem. But if the door is only partially open, a vertical-only array could mistakenly read it as being fully open. Pairing any one of our vertical column of feelers with a mate, and moving the pair from center-forward to the edges of the cylinder, pointing tangentially now, rather than radially (in other terms, arranging that they point in the direction of forward motion rather than straight out along a radius line) provides us with ample warning of a doorway, window, narrow hallway, and so on.

But an empty house paints a bleak picture. If we slowly add items to it, we may be able to determine that only one or two of the feelers need ever be paired, rather than all three. Since small things seem to be the most meddlesome obstacles, we can start with something like a standing ash tray, a stereo speaker, a planter, or a pussycat.

These items are all small enough to be missed by the upper two feelers and narrow enough to be often missed by a central floor-level feeler. Yes, the floor feeler will have to be a pair, which means the lowest vertical areas also must be scanned horizontally across the android's width.

There are traps, it turns out, for the center and upper sections, too: narrow passages between furniture with spindly legs; shelves and fixtures jutting out of walls; narrow chandeliers, fixtures, and so on extending down from ceilings.

If we were to proceed further with this investigation, we would find a need to increase our coverage—either by compromises in its resolution or extensions in its implementation—to cover the entire area likely to be next occupied by the android.

Once we remove our assumption of forward motion, the need for coverage increases again, again *to cover the entire space likely to be next occupied by the android*. And (as will be explained in more detail later) as we include probability shells, we must allow not only for objects *already* in that space, but for objects that *may possibly enter* that space as well.

We have discovered that our "blindfold" experiment is not likely to succeed without some substitute for vision—which is not to exclude vision as a possibility.

Nonetheless, we've made substantial progress. We've seen how the human living environment influences and limits the eventual dimensions of an android. We've seen the nature of the problem to soon face us in designing our collision-avoidance subsystems. We've been introduced to the concept of mapping an area and had just a hint of how probability shells will enter the mapping problem. We've recognized that the manipulative imperative allows us ways around or through otherwise insoluble obstacle-avoidance problems.

But we've left out one particularly sticky problem that's going to make a considerable mark on the eventual mechanics of our android: stairways.

6
The Special Problems of Stairways

Toddlers and the physically handicapped have an especially good understanding of the problems a simple stairway can present. This jagged ramp of precipices poses problems in balance and manipulation even for humans, though a lifetime of practice has made most of us capable of handling them with only an occasional stumble.

We, with our two legs, have learned to conquer stairs. Eventually, we may be able to design an android sophisticated enough, both mechanically and electrically, to emulate two-legged stair climbing. But consider what the task involves.

From a standing or walking start, balance on one foot while leaning toward the top of the stairway; raise or lower your free foot (as appropriate, depending if you're heading up or down the stairs) to the top of the first step; shift your entire body's balance smoothly both forward and laterally until it is on that foot; repeat the procedure until the final step; then readjust your whole body balance and lean for level walking. Biped walking itself has been called controlled successive falling—the problem is amplified by nonlevel surfaces.

Choosing a drive mechanism capable of handling stairs is only part of the problem. We are going to have to very carefully investigate the nature of the stairway problem and the requirements it imposes on our android design. Because gravity is involved, we might anticipate speed, weight, power, and control to be parts of the problem and its answers.

Some of you may be disagreeing already, thinking that walking up and down in emulation of human stair climbing is the obvious answer and that the problems are not insurmountable.

That may be true—given time, money and a lot of attention. But try a little experiment before you decide: walk the stairs the way an android must, one operation at a time. Proceed in *extremely* slow motion. Pay close attention to the changes in position each foot goes through, each toe, each calf, each ankle, each knee, each thigh. Feel your center of balance change from side to side. Notice how you lean forward. Then consider the motors, joints, drives, balance mechanisms, and mechan-

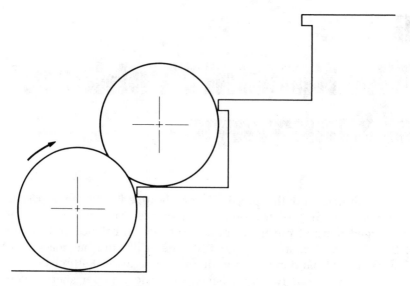

Fig. 6-1 *If wheel diameter is less than twice the rise of a step, wheel can neither mount first step nor easily crest corners of stairs.*

ics, not to mention the speed and complexity of the electronics. If you still think you can make it happen, then, by all means, make it happen!

But two-legged beasties may not be the end of the trail. Dogs and cats with their four legs climb stairs handily—and you and I have been known to crawl them in weaker moments. Four legs greatly simplify the balance problem, but they don't eliminate it. The problem now is one of coordinating balance, alternately, on the front and rear pairs of legs. Again, we face the problem of driving the full mass of the android while only half of its drive mechanism is engaged at any instant; this is compounded by complications associated with gravity—either because of the power required to climb against it or because of the braking required to keep from falling without control.

Six-legged mechanisms have been successfully built that handle stairs easily in both directions. These beasties are spiderlike, with three legs on each side. Since four legs are planted at any one time and only two move, balance is a minor problem and drive and braking difficulties are minimized.

But there is a snag. While the six-legged mechanism is thoroughly at home on a stairway, it shares a problem endemic to all walking mechanisms: it's hard to turn, especially where tight turns are required, as in doorways, at landings, in halls, and so on.

Wheels are another alternative worth investigating. The first variable we should look into is the size of the wheel. Obviously, if a wheel isn't more than twice the size of a step (diameter twice the rise, e.g., more than 16 inches in diameter for an 8-inch tall step), it'll never get up onto the first step (see Figs. 6-1 and 6-2).

In fact, you can determine for yourself that a vehicle with wheels with a diameter of the same order of magnitude as the rise of a step cannot negotiate a stairway in a controlled manner, even if the first-step problem could be overcome. (Only the hardy and foolhardy will actually perform this experiment; the rest of us can be satisfied of its veracity through more intuitive insights.) First, ride a bicycle up and down a short flight of stairs. Then try doing the same in a go-cart. (Ouch!)

Unfortunately, a large wheel usually (but not always) means a high center of gravity, which means stability on a stairway could become a problem area. Unfortunately, as soon as we make a wheel big enough to climb a stairway, it's just big enough to place its hub vertically over a place that's off the edge of the step, giving the large wheel a definite tendency to roll down hill. If you bicycle or remember bicycling, you may recall how much more difficult it is to pedal up a stairway than up an equally steep hill or ramp. (See Fig. 6-3.)

Like walking mechanisms, tight turns are a problem for most wheeled arrangements with the wheel size we require (twice the rise of a step, or about 16 inches in diameter). In fact, only one wheeled arrangement this size permits an android dimensioned within our set limitations to make turns tight enough to get itself through doorways along narrow halls: a tricycle.

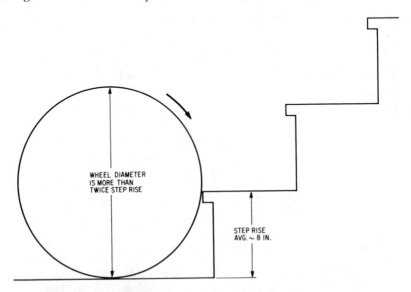

WHEEL DIAMETER
IS MORE THAN
TWICE STEP RISE

STEP RISE
AVG. ~ 8 IN.

Fig. 6-2 *If wheel diameter is more than twice the rise of a step, wheel mounts stairs easily.*

Not even all versions of a tricycle apply. Versions where all three wheels are steerable or even just two out of the three require too much width with these large wheels. Even in the remaining version—one steerable wheel—a cutout in the android body must be made to allow for all possible wheel positions. This further complicates the goal of a low center of gravity for good balance and control.

Also, some mechanism for attitude (degrees out of vertical) control must be incorporated to keep the beastie from tumbling head over motors down the stairs. If we make one concession, we might be able to bring this tricycle design a little closer to feasibility.

Let's assume that the android can back down the stairs. And let's call the face with the two fixed wheels the front, the face with the steerable wheel the back. This permits us to configure the body with most of the mass at the front, hopefully as low as possible. In Fig. 6-4 the single steerable rear center wheel requires spherical clearance in the android body but permits on-wheel mounting of motor(s) and brake(s). The rear wheel is mounted on an elevator mechanism that permits it to drop 8 to 12 inches on stairways and ramps: this permits the android

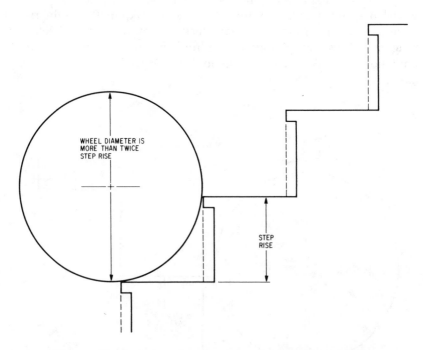

WHEEL DIAMETER IS MORE THAN TWICE STEP RISE

STEP RISE

Fig. 6-3 *Wheel is large enough to easily mount first step and easily crests top step corners. This is true both for flush and kick-riser front geometry. But wheel center is not over contact point— it will always tend to force wheel down stairs.*

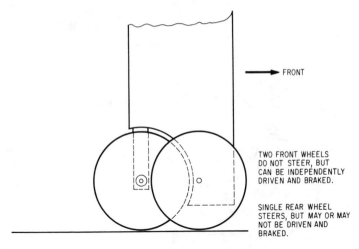

FRONT

TWO FRONT WHEELS
DO NOT STEER, BUT
CAN BE INDEPENDENTLY
DRIVEN AND BRAKED.

SINGLE REAR WHEEL
STEERS, BUT MAY OR MAY
NOT BE DRIVEN AND
BRAKED.

Fig. 6-4 *Side view of tricycle drive wheel configuration on flat surface.*

body to stay level or tilt slightly uphill, aiding control and somewhat compensating for a high center of gravity.

Now, dropping the rear wheel on a flat surface tilts the front of the android down toward the floor by 45° or so; on a 45° incline, though, it lets the android travel upright. Assuming we can build a sufficiently clever attitude-sensing and control circuit mechanism, we may be able to assume, for the time being, that the problem of leaning is solved.

Let's examine android stair climbing in more detail (see Fig. 6-5). The sequence of events to mount the stairway is:

1. Front wheels come to the edge of the stairs with front pointing in upstairs direction. This is true for either climb or descent, since the android must "back" down stairs.
2. Stop.
3. Extend the rear wheel elevator, tilting the android forward about 45°.
4. Slowly mount the first step, bringing the android upright, but slowly, so it does not swing into fall.
5. Climb or descend the stairs.
6. When on a level surface, retract the elevator to right the android.

However, the beastie still falls down the stairs, or at best totters. The reason is that the lowest point on each wheel is still always over the edge of the step.

Big wheels seem a bad round peg for a stair's square notch. But there is a way of making "synthetic" big wheels out of smaller ones.

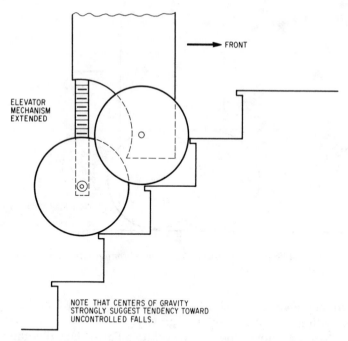

ELEVATOR
MECHANISM
EXTENDED

FRONT

NOTE THAT CENTERS OF GRAVITY
STRONGLY SUGGEST TENDENCY TOWARD
UNCONTROLLED FALLS.

Fig. 6-5 *Side view of tricycle drive wheel configuration on stairs.*

Back in the sixties, when lunar excursions were still in the future and NASA's budgets were still sizable, the public was often treated (through the media) to artists' renderings of some fantastic but feasible schemes for moon buggies. They walked, they crawled, they slithered and rolled, and more than anything they kept us as excited about the future as any of the Gernsback science fiction covers of the thirties and forties did our fathers. One of those renderings came to mind when the problem of climbing stairs came up. While the rendering itself is not available, memories of it may describe the scheme in enough detail to entice some of you into building and experimenting with it.

Three wheels are used, with their centers at the points of an equilateral triangle; the wheels do not touch each other. A point determined to be at the triple junction of lines bisecting the three angles of the triangle is the center of the fourth, smaller wheel. This fourth wheel is the drive wheel for the entire mechanism (see Fig. 6-6).

The hubs of the three outer wheels are sprocketed, as is the outer rim of the fourth, inner wheel. The inner wheel diameter is chosen such that it touches a chain connecting the three outer hubs at three points, each midway between two adjacent hubs. The three outer wheels are chain driven by this fourth inner wheel in unique pseudo-planetary drive.

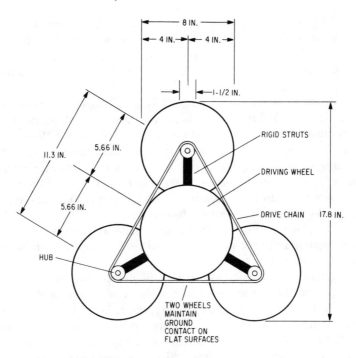

Fig. 6-6 *Triangular wheel drive.*

With this drive, two wheels normally stay in contact with the ground for travel over relatively smooth surfaces. Once the leading wheel meets an impasse, though, the drive treats its hub as the fulcrum of a lever and the remaining wheels leapfrog it. On a stairway, successive leapfrogging easily brings the mechanism upstairs; control during descent, however, remains a ticklish problem.

The dimensions for this wheel drive shown in Fig. 6-6 are for standard 8-inch steps. Otherwise:

Step rise ≈ step tread
Let x = step rise
Wheel diameter = x
Wheel center spacing = $x\sqrt{2}$
Overall height = $x(1+\sqrt{3/2})$
Between wheel spacing = $x(\sqrt{2} - 1)$
Minimum driving wheel diameter = $x\sqrt{2/3}$
Driving wheel diameter = $x\sqrt{2/3}$ + hub diameter

Tight turns can be accomplished by driving one side forward and the other backward. A low center of gravity is possible. All in all, the configuration shows promise, and you are encouraged to pursue it.

The one disadvantage of the scheme is that there has not been a great deal of experimentation and development done with it, so we are very much on our own in pursuing it. There is yet one more approach that has had nearly a century of commercial development: track drive.

Properly designed, a tracked vehicle can easily overcome the problems a stairway brings. An inverted trapezoid configuration assures good posture for mounting the first step on ascent and allows the upward tilt to happen in two phases: first, through the ramping effect of the leading edge of the track, then through the actual ramp the stairway provides. If the floor-contacting portion of track is long enough to span the top corners of three steps, continuous contact with at least two steps through all parts of an ascent or descent is assured.

A vehicle with twin tracks and separate reversible drives for them is capable not only of forward and reverse locomotion but also of extremely tight-radius turns.

There still remains the problem of toppling, but with careful design this can be overcome. This requires, of course, a low center of gravity. A track drive allows us ample room to place the android's most massive components—batteries, motors, and so forth—very close to the ground. In addition, we can resolve to keep the upper parts of the android as lightweight as possible. Also, we ought to consider a leaning mechanism.

The special problem of stairways seems to lead us to the special circumstance of a track-drive machine. Let's continue on the basis that while we will be designing and depicting a track-drive machine, little of the remaining design requires a track drive. In any case, while specific responses may change, our problem/solution approach to design will not.

7
Designing the Main Mechanical Drive

We know about several of the dimensional requirements for our drive mechanism—but we'll need to tighten them down a bit. In this chapter, we'll design a main drive chassis and subchassis, a track drive to go with it, and a triangular wheel drive we can substitute for the track on the same chassis.

We also are going to depart a bit from the step-by-step process of previous chapters; instead, I'm going to share the conclusions I've reached in my own investigations and the rationale behind them. Again,

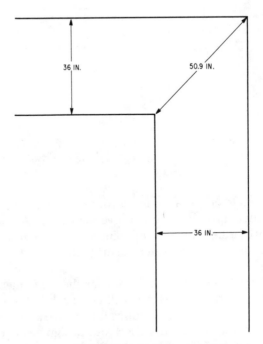

Fig. 7-1 *A 90° turn in narrow 36-in. hallway represents important common requirement for drive mechanism. Note that right and left turns are both shown here, depending on approach.*

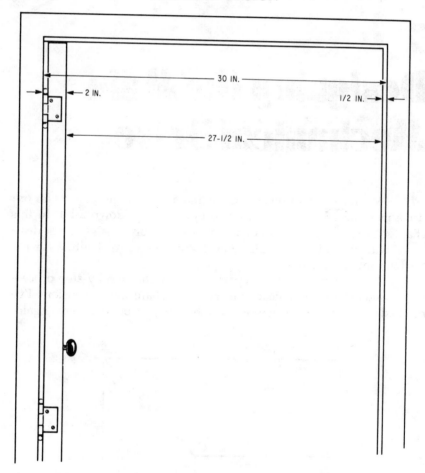

Fig. 7-2 *Even a 30-in. doorway only offers about 27 in. of free space because of intrusions of door edge and molding into available width.*

you are welcome and encouraged to do your own footwork, to disagree, to take a different approach, even to prove me wrong. At the very least, I hope this proves a helpful starting point for your own thinking.

As a starting point, consider the drive chassis to be a rectangular solid. How big should it be?

One of the worst-case conditions for the drive mechanism is a 90° turn in a narrow hallway, which will occur when the hallway presents a corner or a doorway that requires this turn. Since most hallways are at least 3 feet wide, we should check our design against the dimensions of such a turn (see Fig. 7-1 on the previous page).

Before doing so, we can respond to the width of the doorway itself by setting the maximum width of the mechanism at 27 inches. This corresponds to a 30-inch doorway with an open door edge occupying

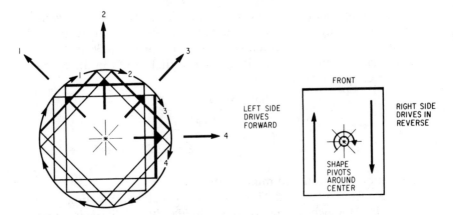

Fig. 7-3 *Driving one side forward and opposite side in reverse causes mechanism to pivot in circle with radius equal to half the rectangle's diagonal and diameter equal to diagonal.*

2 inches of the doorway, molding occupying 1/2 inch, and the last 1/2 inch accounted for by a very minimal safety margin (see Fig. 7-2).

The minimum turning radius of a wheeled or tracked mechanism occurs when one side is driven forward, the opposite side in reverse. This causes the mechanism to pivot about its horizontal center (see Fig. 7-3).

The corner of a 36-inch hallway right angle turn features a maximum dimension of 51 inches, diagonal corner to corner—but this is a misleading dimension. As shown in Fig. 7-4, the circle A diameter is

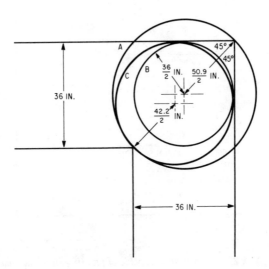

Fig. 7-4 *Pivot circles in corner of hallway right angle.*

the corner diagonal (50.9 inches), but the circle fits inside the walls of the hallway at only four points—the four corners. Not surprisingly, the circle B diameter wholly contained inside the walls is 36 inches—the hallway width. Circle C is the largest circle that can fit in the corner: it is the easiest pivot circle.

$$\text{Diameter} = \frac{\text{hallway width} \times 2}{1 + \sin 45°}$$

$$= \frac{36 \text{ inches}}{1.707} \times 2 = 42.2 \text{ inches}$$

The position is critical and must be precise.

A rectangular solid 27 inches wide capable of pivoting in circle C would be 23.8 inches long, as shown in Fig. 7-5. We'll see in a minute that this is not long enough for a mechanism intended to climb stairs.

It turns out that the minimum turning radius without regard to walls is not appropriate here because the corner, while square, will accommodate a 90° arc of a wider circle. This is actually a family of circles with a difference between inner and outer radii of 27 inches.

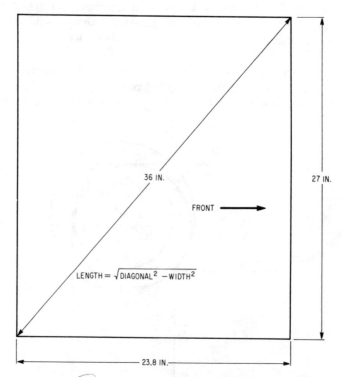

36 IN.

27 IN.

FRONT ⟶

$$\text{LENGTH} = \sqrt{\text{DIAGONAL}^2 - \text{WIDTH}^2}$$

23.8 IN.

Fig. 7-5 *A 27-in.-wide drive capable of pivoting in 36-in.-wide circle is 23.8 in. long.*

Circle B, the minimum inner circle (see Fig. 7-4), is the one of zero radius, corresponding to a turn that begins with the rectangle hugging the inner wall. This also corresponds to a turn in which the track along this wall stops (and slides through rotation) as the outer track continues. The rectangle proceeds until halfway through the corner before the turn begins. This movement is more precisely shown in Fig. 7-6. By leaving the inside track stationary while driving the outside track, the drive pivots about a point at the center of the inside track, resulting in the progression shown. For maximum length:

$$\text{Length} = 2 \times \sqrt{(\text{hallway width})^2 - (\text{drive width})^2}$$

where

$$\text{Turning radius} = \text{hallway width}$$

The maximum length can be then determined by assuming the width as 27 inches, and the maximum distance from the center of one side to an opposite corner to be the hallway width, 36 inches. This yields a maximum length of 47.6 inches. At this nearly 4-foot length, the rectangle just misses all the walls.

The minimum permissible length for the rectangle is determined by the stair-climbing requirement (see Fig. 7-7). In order to assure proper control and traction while climbing or descending stairs, the drive should contact at least two steps at all times, requiring an end-to-

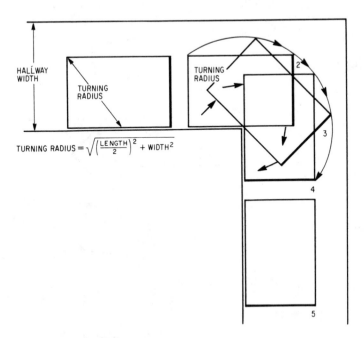

Fig. 7-6 *Progression of rectangular solid turning corner of hallway right angle.*

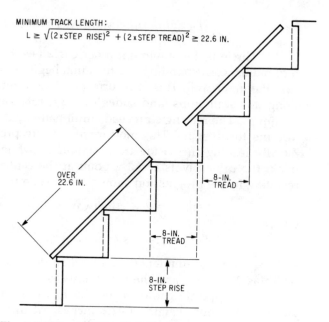

MINIMUM TRACK LENGTH:

$$L \geq \sqrt{(2 \times STEP\ RISE)^2 + (2 \times STEP\ TREAD)^2} \geq 22.6\ IN.$$

OVER
22.6 IN.

8-IN.
TREAD

8-IN.
TREAD

8-IN.
STEP RISE

Fig. 7-7 *Using stairway dimensions to determine minimum grounded contact length of track. Note: It makes no difference whether the stair is square or lipped.*

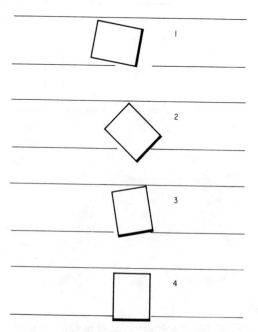

1

2

3

4

Fig. 7-8 *Maneuvering a shape in a hallway.*

end length that allows it to sit atop three steps. With standard 8-inch steps, this requires twice 11.3 inches, or 22.6 inches. This size can turn in a 29.3-inch hallway.

But, getting back to defining a maximum, the most difficult case involves turning from a 36-inch hallway into a 27-inch doorway, and vice versa. This may require a number of repositioning moves, all of which involve the rectangle at one point facing the doorway at right angles to the hall. Obviously, the 36-inch hallway dimension now becomes even more significant, and, in fact, sets our maximum length. This size can turn in a 32½-inch hallway corner in a single move, permitting a narrow error margin in traversing the hall.

As shown in Fig. 7-8, note that while it may be possible to maneuver a shape as long as the hallway is wide and as wide as the doorway, it requires extensive maneuvering. In any case, the task is impossible if the length exceeds the hallway width, unless the shape is extremely narrow.

Things get easier, of course, as we make the track chassis narrower (see Fig. 7-9). This permits tighter turns, more room for error, passage through more constricted paths, and lightens the weight of the beastie overall. My own design is based on 20-inch width.

We know the minimum and maximum lengths, in the end, are affected by hallway and step dimensions more so than by the corner situation we began with. How compatible are these dimensions with the stairway-mounting problem?

Tracked mechanisms generally mount stairs by ramming into them; the first surface to meet the stairs is angled so that the ramming action forces a climbing action (see Fig. 7-10). We can help this climbing action by placing the top track idler pulley at a height that puts its center above the lip of the step; it then will tend to roll over the leading edge

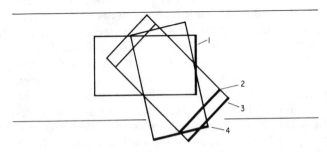

Fig. 7-9 *When shape is narrowed, maneuvering is simplified. To move from position 1 to position 2, right track remains stationary while left track goes forward; from position 2 to 3, both go forward; and from position 3 to 4, right track goes in reverse while left track goes forward.*

of the first step (see Fig. 7-11). Since most steps are 8-inches tall (8-inch rise), we can set this dimension at 9 inches.

We have now described an inverted trapezoidal solid, assuming front/back symmetry, as shown in Fig. 7-12. This is a helpful assumption because it means that there is no difference to the android whether it is going frontward or backward at any time, which reduces the amount of energy it has to expend in positioning itself for stair climbing or any other task.

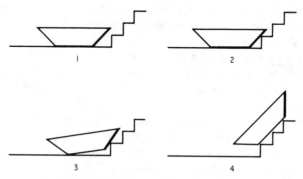

I.   TRACK APPROACHES STAIRS.

2.   LEADING SLANTED EDGE TRANSLATES RAMMING FORWARD
     MOTION INTO WEDGING CLIMBING MOTION.

3.   CLIMB CONTINUES. SOME STRESS LIKELY SINCE RAMMING
     AND CLIMBING HAPPEN AT DIFFERENT SPEEDS. ELASTIC AND
     TENSION-ABSORBING TRACK ELEMENTS MUST ABSORB
     DIFFERENCE.

4.   ONCE MOUNTED ON STAIRWAY AND OFF LEVEL FLOOR
     APPROACH, STAIRWAY CAN BE MODELED AS BUMPY RAMP.

Fig. 7-10 *Track approaching and climbing stairs.*

We can proceed at this point by making a few assumptions that may (and in fact probably will) later prove false. One is that we can dictate the sizes of the mechanical components and eventually find a close enough commercial part for each guessed-at piece. Another assumption is that you will not regard the plans detailed here as either definitive or final but rather will work through your own design from scratch. A third assumption is that somewhere, sometime someone will prove everything we try impossible, even if it works. Examples of this latter assumption will be incorporated in this text as they are discovered.

The height of the carriage, as we've seen, depends on the height (rise) of a step and the diameter of the track-top wheel. But before we go much farther, it's time to define a few terms—or at least a few names—for the components in the track mechanism.

Track refers to the connected or continuous elements that are driven in contact with the floor, ground, or other surface in order to

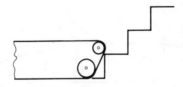

Fig. 7-11 *Arranging track drive configuration so that top pulley rolls over and onto top corner of first step aids climbing aptitude.*

propel the vehicle forward. Tanks and bulldozers, for example, are tracked vehicles. Figure 7-13 identifies parts of the track mechanism.

Generally, tracks are bands or belts (continuous or linked-element) wrapped around arrays of *wheels*. These wheels may take the form of large, toothed gears, and are then called *sprockets*. Or they may take the form of flanged wheels with teeth (or without) and are then called *pulleys*.

Some number of these wheels or sprockets or pulleys (as few as one, as many as all) will be directly or indirectly connected to the propelling motor or motors. These are referred to as *drive wheels* or *driven wheels* (or sprockets or pulleys). Those that are allowed to freewheel are called *idlers*. A special class of idlers, called *rollers*, includes small load-bearing (and load-sharing) wheels mechanically part of the vehicle carriage, located and positioned to ride atop the ground-contacting portion of the track itself.

Surface features on the ground-contacting surface of a continuous track are called *cleats;* they resemble ribs and may be a molded part of the belt or added on at any time.

Linked-belt tracks generally do not contact the ground directly. Instead, a *shoe* is attached to each link. Cleatlike bars on these shoes are

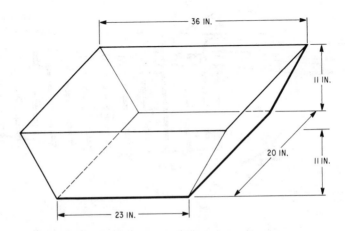

Fig. 7-12 *Drive can be modeled as a trapezoidal solid.*

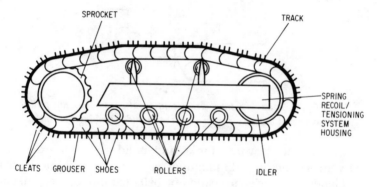

Fig. 7-13 *Track mechanism.*

called *grousers.* Cleats and grousers on tracks do the job that tread features perform for tires: they increase the surface area of the track and improve its tractive abilities; in climbing modes, they improve *toeing—* the track's grip or handhold on surface features.

The distance between the center of the two tracks (in a two-track drive, which is the most common configuration) is cabled the *track gauge.* For best maneuverability and optimum steering characteristics, the track gauge should be at least 7/10 of the grounded contact length. For our 23-inch grounded contact length, this suggests a minimum 16.1-inch track gauge.

Tracks may be constructed of the continuous toothed high-torque drive belts now replacing chains and v-belts in many applications in industry (see Fig. 7-14). Uniroyal, one of the leading manufacturers

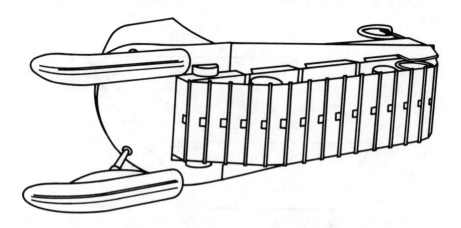

Fig. 7-14 *Snowmobile track is special high-torque drive belt. (Courtesy of Uniroyal)*

of these belts, offers a number of design aids for would-be users of these belts, some of which will be detailed later.

Of immediate concern are some of the terms used in referring to or specifying these belts (see Fig. 7-15).

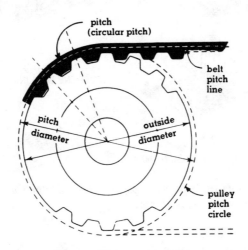

Fig. 7-15 *Definitions of terms used in describing and specifying high-torque toothed drive belts. (Courtesy of Uniroyal)*

The *pitch* of one of these belts is defined as the distance between the centers of two adjacent teeth, measured along the *pitch line* of the belt; an identical measurement would be made if the belt were cut and made to lie flat. The *pitch line* of the belt is an imaginary line that extends along its entire length, located halfway through the "skinny part," the part of the belt that would remain if all the teeth were ground flat. This "skinny part" of the belt is its *tension member*.

The belt *pitch length*, or circumference, is its total length as measured along the pitch line. Pitch length, pitch, and width are the three principal dimensions of a belt of this variety.

The flanged sprocket pulleys (yes, all three terms at once) made to mate with these belts are principally described, dimensionally, in terms of the *number of grooves*, *pitch*, and *width*. The *pitch diameter* of the sprocket, which is always greater than the face (or outside) diameter, is measured against its *pitch circle*; the pitch circle of the sprocket coincides with the pitch line of its mating belt when in place around it. (See Fig. 7-16.) For now, we'll ignore flange and bushing and bore dimensions, but these are all part of the final specification.

A virtually unknown track configuration is the Darragh mechanism. Here, the track itself is neither flexible nor in contact with the

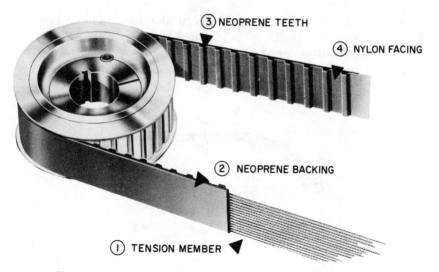

③ NEOPRENE TEETH

④ NYLON FACING

② NEOPRENE BACKING

① TENSION MEMBER

Fig. 7-16 *Construction of Uniroyal HTD toothed belting, also called timing belt. (Courtesy of Uniroyal)*

ground. Instead, it is a rigid race (shaped, no pun intended, like a racetrack). Shoe elements are mounted about its circumference on pairs of tapered roller bearings, one pair on the outside of the race, a second pair (more narrowly spaced) on the inside; the angles between outer and inner centers versus perpendicular are ± 15° (see Figs. 7-17 and 7-18).

Links join the left inside bearing of one shoe with the right inside bearing of its neighbor to the left, on around the track. The configuration is such that all rollers are in intimate contact with the track

Fig. 7-17 *Three views of early working prototype of Darragh tractor tread mechanism, circa 1959.*

NEW TRACTOR TREAD

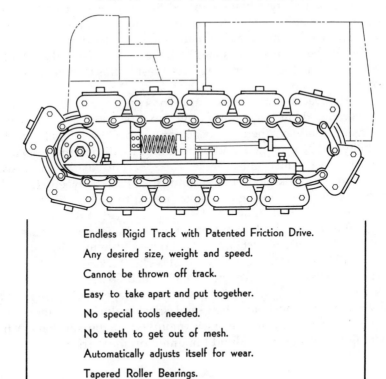

Endless Rigid Track with Patented Friction Drive.

Any desired size, weight and speed.

Cannot be thrown off track.

Easy to take apart and put together.

No special tools needed.

No teeth to get out of mesh.

Automatically adjusts itself for wear.

Tapered Roller Bearings.

Better traction, smoother riding and easier steering.

DARRAGH MECHANISMS INC.

3239 FOURTH AVENUE   PHONE: B.F. 6237-J   BEAVER FALLS, PA.

Fig. 7-18 *Old promotional flier for Darragh Mechanisms Inc., since gone out of business. Apparently, Patent 2897014 was issued for this unique track configuration, possibly in 1959.*

at all times, permitting all shoes to slide freely around the track, but permitting no play.

The drive wheel occupies one or both circular ends of the inside area of the track, butting a hard edge against the inside surface of the inside rollers (the outside surface being the one in contact with the track). Stainless steel or other hard steels are the materials of choice. The drive wheel drives the inside bearings, which propel the linked shoes around the track; when the assembly is placed in contact with a stationary surface beneath, this drives the whole thing forward.

The most significant features of the Darragh mechanism are its mechanical simplicity when compared to sprocket drives and its construction from nonelastic materials. Also, this mechanism is especially well suited to climbing because the unique geometry of the offset inside/outside members causes the shoes to move around the ends faster than they move along the straightaways. As a result, this track tends to climb faster than it rams.

The Darragh mechanism also has two inherent disadvantages. First, because there is no control over specific shoe position as it approaches a stairway, there is no guarantee that the inherent climbing capability can be exploited, and there is little predicting the exact failure mode should the mechanism be jammed. Second, as the shoes swing down and into rank along the ground, the space between shoes then closes like pincers, causing a definite hazard to puppy dog tails, lamp cords, feet, and other things we love that may get in its way.

But let's get back to our own track carriage, now that we know what to call things. In fact, we can even be a little more specific about the overall track drive mechanism parts. The track itself and the track mechanism (reduction drive, etc.) are mounted on a *subchassis;* a *carriage chassis* nests within it. The subchassis and carriage chassis are mechanically joined by shock-absorbing and shock-reducing components to keep jolts originating in the track (mounted to the subchassis) from reaching the rest of the android (mounted on the carriage chassis).

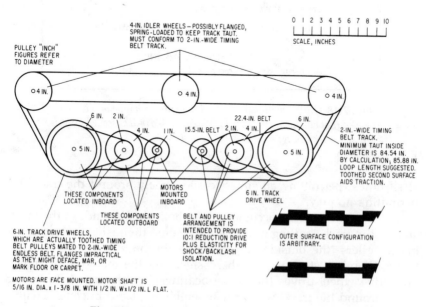

Fig. 7-19 *Identification of track drive mechanism.*

In determining the proper height of approach to a step for the top of a track, the significant dimensions are the vertical and horizontal offsets of the wheel (or pulley or sprocket) centers and their diameters. Except for Darragh mechanisms, the actual track belt width and thickness are not very significant for this calculation.

To keep flexing on the track reasonable, it's important to ensure some minimum size for the curves it has to wrap around. In dealing with toothed drive belts, a suitable minimum pitch diameter for track mechanism pulleys is about 4 inches (see Fig. 7-19).

For the inverted solid trapezoidal shape we are modeling this track mechanism around, there are compelling reasons for making the top pulleys *idlers* and the bottom pulleys *drive pulleys*. One is the requirement that we maintain as low a center of gravity as possible (for optimum stability, especially when climbing); driving the bottom pulleys permits placing the motors and reduction mechanism (rotational speed reduction, assuming it is needed, between the motor and the drive pulley) at the bottom of the subchassis. Another is the ease of incorporating spring tensioning in the idler mounts if they are not load-bearing elements. Also, placing the drive pulleys only a track-belt-thickness away from the ground permits intimate driving contact, meaning more efficient power transmission for propulsion because of fewer losses through elastic absorption in the track itself.

Six inches seems a suitable size for the drive pulleys. This permits a pulley small enough to keep weight reasonable and allow adjacently mounted pulley pairs room on the subchassis; these latter pulleys are one approach to a reduction mechanism. Six inches also permits a minimal but ample 1-inch ground clearance for the subchassis while permitting good low mounting positions and clearance for all drive components.

We're finally in a position to determine actual dimensions and positions for the tracks and pulleys (see Fig. 7-20). We wanted to put the center of the top (idler, 4-inch) pulley 9 inches above the floor to assure a good roll-over action as the track mounts the first step of a stairway. That positions its center 6 inches higher than the center of the drive pulley (6-inch diameter), which is 3 inches above the floor.

We also can determine that the outside idlers must be horizontally positioned with their centers 32 inches apart to meet our 36-inch overall length restriction.

The grounded contact length of the track was to be 23 inches, determined by the span between crests on a stairway. Since this length of track must contact the ground, this 23 inches also is the horizontal distance between centers for the drive pulleys.

We now have enough information to calculate the actual angles, arcs, and distances involved in determining the overall track length.

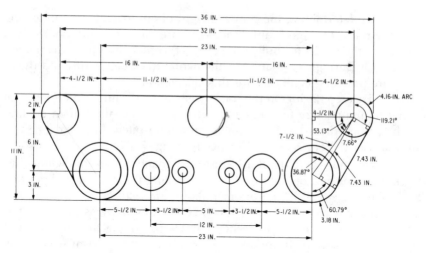

Fig. 7-20 *Dimensions of track drive mechanism.*

There are four straight sections of track and four curved sections. The top straight section is identical to the horizontal distance between the outside idler pulley centers, which is 32 inches. Similarly, the bottom straight section is equal to the distance between the drive pulley centers, or 23 inches.

The straight slanted front and back sections can be calculated on the basis of two triangles. The first triangle is made of a vertical line extending straight up from the center of a drive pulley to the line between top idler pulley centers, which it intersects at a right angle; a segment of that top line extending from this intersection to the near outside idler pulley center; and a straight line between these drive and idler pulleys.

This last line length is determined by the Pythagorean theorem as the square root of the sum of the squares of the first two lines. The vertical element is the height of the idler center above ground (9 inches) less the radius of the drive pulley (3 inches), or 6 inches. The horizontal line length is equal to half the difference of the distance between outer idler centers (32 inches) and drive pulley center separation distance (23 inches), or 4-1/2 inches. This yields 7-1/2 inches as the distance between pulley centers.

The second triangle is described by this line; by a line parallel to (and equal in length to) the straight section of track; and by a section of the radius of the drive pulley extending from its center toward the point on its circumference where the straight and curved track portions meet for a distance equal to the difference in radii of the drive and idler pulleys. The end of this latter radial segment meets the previous (parallel to track) segment at a right angle; the other end of the parallel segment is then located at the center of the idler pulley. (See Fig. 7-20 for clarification.)

Pythagoras is called on again to yield a straight end run of track equal to the square root of the difference of squares of the distance between pulley centers (7-1/2 inches) and the radial (difference of radii) segment (1 inch), or 7.43 inches.

At this point, we will need to calculate the angles within this triangle and soon those within the first triangle as well, in order to determine the curved track dimension, which occupies four arcs along the circumferences of the four corner pulleys. These arcs are calculated by determining what proportion of the full 360° circumference of each pulley is occupied by the track.

For this last triangle, the narrow top angle converging at the idler center is the angle whose sine is the ratio of the short (difference of radii) opposite side to the longest (distance between centers) side, or arc sin 1/7.5, equal to 7.66°. The wider angle converging at the drive pulley center can be calculated either as the arc cosine of this ratio or as the difference of the first angle and 90°. Either way, the result is 82.34°. A similar calculation for the first triangle yields the arc sin of 6/7.5 as 53.13°, the arc cos of this ratio as 36.87°.

Now, examining the track arc along the drive wheel is the difference between 180° and the sum of the triangle angles converging at its center, or 180° − (82.34° + 36.87°) = 180° − 119.21° = 60.79°. The arc is then $\pi \times 6 \times (60.79/360) = 3.18$ inches per pulley.

The arc for the top pulley may be calculated in the same way, but there is a significant shortcut available. Notice that the total arc of the track from bottom to top is 180°; the angle of the arc at the top, then, is the difference between 180° and the angle at the bottom, 180° − 60.79°, or 119.21°. The top arc is then $\pi \times 4 \times (119.21/360) = 4.16$ inches per pulley.

Each end, then, requires a track length equal to the sum of the straight portion and the two arcs, 7.43 inches + 3.18 inches + 4.16 inches, or 14.77 inches.

Finally, the total track length is the sum of the top plus the bottom plus the two ends, or 32 + 23 + 2 × 14.77 inches = 84.54 inches.

That, at last, is how track length calculations are done and what the results for our model come out to. But we are still dealing with a model only. How does the real world change our view of the track mechanism? (It cost me a lunch with an expert from Uniroyal to find out.)

I'll try to spare you from the rigorous chapter-and-verse approach of the previous exercise. The result of a thorough investigation of the Uniroyal Industrial Products Power Transmission Division (Middlebury, CT 06749) *Timing Belt Drives* manual yields the following choices for our track drive:

Belt Model 850H. 170 teeth. Total pitch length 85.00 inches. 1/2-inch pitch. Model 850H200. Belt width 2 inches. Approximate weight 1.32 pounds.

Drive Pulley Model 36H200. For belts 2 inches wide, 1/2-inch pitch. 5.730-inch pitch diameter. 5-61/64-inch flange diameter. Approximate weight 8 pounds. Requires Model QD/SK *Bushing,* approximate weight 2 pounds.

Idler Pulley Model TL26H200. For belts 2 inches wide, 1/2-inch pitch. 4.138-inch pitch diameter. 4-3/8-inch flange diameter. Approximate weight 3.8 pounds. Requires Model TL2012 *Bushing,* approximate weight 1.4 pounds.

Checking these diameters against the pulley-center-position dimensions of our model, we can calculate an end dimension of 14.74 inches for an overall pitch path length of 84.47 inches. Aha! A close if not perfect fit for our 85-inch belt.

It took me a long and laborious evening with my scientific calculator in hand to determine empirically that the best way to lengthen the path to accommodate the actual belt pitch length is to increase the height of the idler pulley centers. The actual height between the drive pulley centerline and the idler pulley centerline that fits an 85-inch pitch-length belt falls between 6.254 and 6.255 inches. Fortunately, mechanics can help rescue the mathematics from nth-place cruciality.

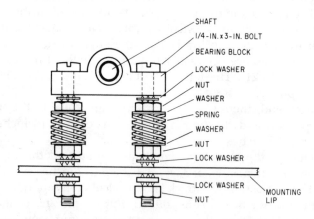

Fig. 7-21 *Mounting of shaft/bearing block using springs for tension and variable position.*

Our rescue comes from a not-very-new modern mechanical contraption—compression springs (see Fig. 7-21). Tom Ryan at Associated Spring was very helpful in providing the design data I needed.

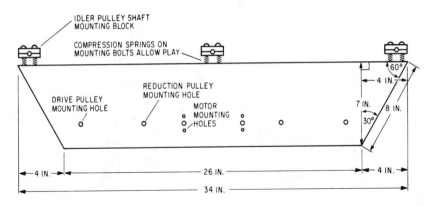

Fig. 7-22 *Main mechanical drive subchassis and side plate.*

I ended up specifying a "C 0600-063-0750-M" oiled music wire compression spring. Its relaxed or free length is about 3/4 inch (0.750 ± 0.020 inch) with an outside diameter of 0.600 ± 0.015 inch. It has a right-hand helix with squared and ground ends. Its maximum *solid* (fully compressed) length is 0.314 inch (2.39:1 compression ratio). It is load rated at 18 ± 1.8 pounds at 0.400 inch, 51 ± 5.1 pounds per inch rate.

This load rating can be multiplied by the number of springs used in parallel. In a mounting configuration where the idler pulley shafts are block mounted with spring-loaded bolts, this means two springs per block (see Fig. 7-22). With three idlers, a total of six springs are on each side; with two idlers, the total is four springs. So, with two to three idlers, 20 to 30 pounds of force can be applied toward the tensioning of each track for every 1/10-inch compression of the springs.

Another view is that the spring mounting provides enough elasticity to absorb 75 to 110 pounds per side, or 150 to 220 pounds overall of forced loading. This provides a mechanism for maintaining reliable track tensioning, responsive to real-world shocks, jolts, and bumps. There will be no track slippage. And adjusting the mounting bolts varies both the position of the idlers versus the track and the total track pitch length while varying the amount of compression on the springs, and thereby tension on the track. All in all, spring mounting of the idler shafts is a neat way of gobbling up lots of little mathematical and mechanical error sources into a few simple adjustments.

A number of spring design and specification aids are available by writing Associated Spring, Barnes Group, Inc., 18 Main Street, Bristol, CT 06010.

So, at least on the surface, we have described a track mechanism that can be built from commercially available parts. The Uniroyal timing

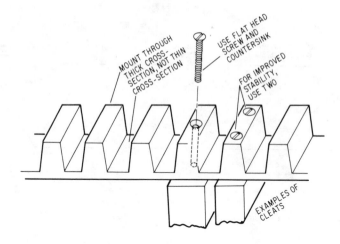

Fig. 7-23 *Timing belt showing how to add cleats.*

belt also is available in a double-sided version for improved traction, but there is another tractive option: cleats (see Fig. 7-23).

The Flexible Steel Lacing Company (Flexco®, 2525 Wisconsin Avenue, Downers Grove, IL 60515) and Tatch-A-Cleat Products® (9460 Telstar Avenue #5, El Monte, CA 91731) offer a series of cleating products that can be cut to size and attached after-the-fact to a timing belt (or any overall-flat belt). When attaching anything to a timing belt, make certain that any staples, screws, rivets, or brads are inserted at the center of the tooth (thick part of belt cross-section); this is the part of the belt-pulley geometry offering the best clearance, as well as the best choice for rigidity and wear. In any case, countersinking of hardware is highly advisable.

We will look at the motor-to-pulley drive later, but there is a motor-related problem we want to look at here. The more the mechanism weighs, the bigger a motor it requires. The bigger the motor, the bigger the battery it requires. And bigger batteries weigh more. Also, the bigger the battery, the more the on-board battery charger will weigh. Then, too, is the problem of finding room for it all. The problem of finding higher-current switching components also is a significant complication.

In short, there are a lot of good reasons to keep the android as lightweight as possible. How well does our track mechanism fit that requirement?

With three idlers, two drive pulleys, and a double-faced belt on each side, our track mechanism weighs in at over 75 pounds! There are ways to reduce its weight, and for those who are planning track drive, they're strongly suggested.

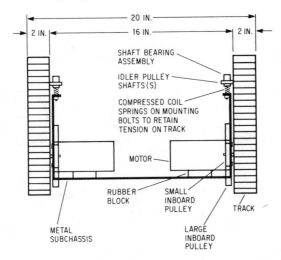

Fig. 7-24 *Front/rear view of track drive mechanism and subchassis.*

First, we can drop more than 10 pounds by reducing the number of idler pulleys on each side from three to two. Second, catalog information on the horsepower ratings of our Uniroyal belt show that the 2-inch width offers more than ten times the horsepower capacity we need.

On a stairway, figuring that each track contacts 1/2 inch of lip contact at the crest of each of two steps, a 2-inch track belt divides the android weight over 4 square inches. For a 100- to 200-pound android, this means 25–50 pounds per square inch—some of which is absorbed by the track, some by the motors, some by the springs in tensile terms. The 2-inch belt is rated for an allowable working tension of 300 pounds.

A 3/4-inch belt is rated at 99 pounds and divides the android weight over 1.5 square inches. A 150-pound android would be stretching (no pun intended) its capabilities.

The weight of the drive mechanism, though, comes down to about 50 pounds with a 3/4-inch set of pulleys.

A 1-inch belt allows a 2-square-inch footprint on the stairs and offers a 144-pound rating, permitting the full 100- to 200-pound weight range while permitting some remaining safety margin. Weight for the mechanism is still under 60 pounds.

Belts other than timing belts could be used, with much more lightweight pulleys, but the very tight control over drive and the precise position feedback with timing belts override the weight savings. (See Figs. 7-24 and 7-25.)

Fortunately, we have an option. We can fall back to the one alternative to track drive that still permits stair climbing: the triangular wheel drive. Better yet, the total weight of four triangle-wheel units should be about 20 pounds.

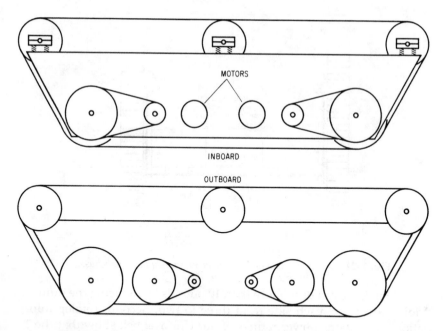

Fig. 7-25 *Track drive mechanism: detail of 10:1 reduction belt and pulley drive.*

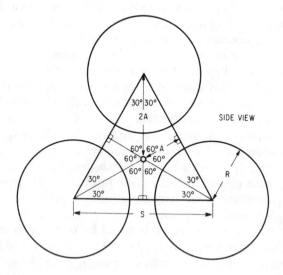

Fig. 7-26 *Basic geometry of triangular wheel drive. Three wheels of radius R are located with their centers at the corners of an equilateral triangle, of dimension S on each side. Unless S > 2R, the wheels will run into each other. The triangle center is located distance 2A from each wheel center.*

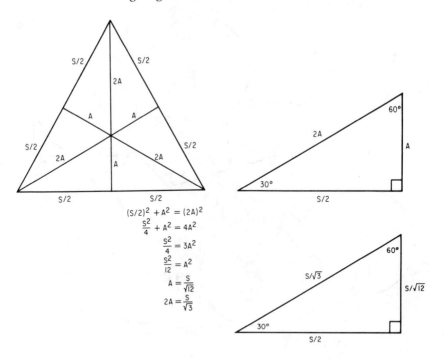

The equations shown in the figure:

$$(S/2)^2 + A^2 = (2A)^2$$
$$\frac{S^2}{4} + A^2 = 4A^2$$
$$\frac{S^2}{4} = 3A^2$$
$$\frac{S^2}{12} = A^2$$
$$A = \frac{S}{\sqrt{12}}$$
$$2A = \frac{S}{\sqrt{3}}$$

Fig. 7-27 *Determining the relationship of A to S.*

The three wheels are located with their centers at the three corners of an equilateral triangle and are locked in rotational step with each other and the drive shaft (see Figs. 7-26 and 7-27).

Since this is an experimental mechanism and the mathematics of the mechanics are an order of magnitude or two more complex than those associated with our track mechanism, experimentation (an empirical approach) is likely to prove preferable to calculation (a mathematical approach). As a compromise, the geometries can be modeled on paper.

The following are the results of my own investigations, which have not yet been tried in hardware.

Clearly, the stair-climbing mode offers more critical, more crucial demands than a run along a flat, level surface (see Fig. 7-28). Here, the difference between lipped and square step geometries is an important one, with the lipped geometry presenting the more difficult problem (see Fig. 7-29).

Consider that, as previously described, the wheels are forced into a leapfrog mode. Mounting the first step is not so much a problem as proceeding from the first to the second. As the leapfrogging occurs and the top wheel assumes the forward position, the rear wheel must

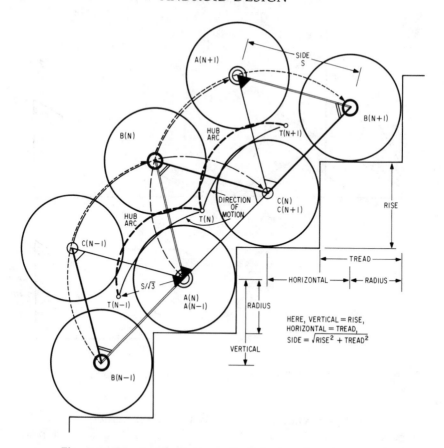

Fig. 7-28 *Progressive climbing action of triangular wheel drive.
Note: Each triangle side, each wheel center, and each angle has its
own stripe pattern for progression up the stairs from time* N − 1
to time N *to time* N + 1. *Wheels are labeled* A, B, C; *triangle center
is labeled* T.

clear the lip on its way to the top position. Complicating this action, we
must consider that before the leapfrogging begins, the rear and forward
wheels had been driven forward until progress was inhibited, the condition that first prompted the leapfrogging. This same difficulty occurs
for the forward wheel on its way to the rearward position.

Previously, we had calculated the linear distance between stair
crests as 11.3 inches for stairs with an 8-inch rise and 8-inch tread. If
we kept each side of the triangle to this 11.3-inch dimension, the leading
edge of a rising wheel would tend to rip the lip off the step during its
arc of rise. Even given an indestructible step, the contact with the lip
would tend to force the rising rear wheel backwards. If successful, it

could bring the lead wheel back to a point where the leapfrogging would not be initiated, and the whole thing would bounce back and forth like a dribbling ball. Even if this didn't happen, opposing stresses could be harmful to the linking drive.

A solution can be found by lengthening the sides of the triangle to 12 inches (see Fig. 7-30). Since the step rise is fixed at 8 inches (the tread is not a limiting factor), the horizontal separation of the wheel centers comes out as 8-15/16 inches.

As a result, as the top wheel rotates into the front position, its arc carries it into the vertical riser. Note that during the leapfrogging, the rotational motion of the wheels has been translated into the rotational motion of the assembly, so the wheel is turning at exactly the same speed as it is falling (none at all vis-à-vis the triangle). Also, once this contact with the vertical riser has occurred, the mechanism is balanced so that gravity will irrevocably complete the leapfrogging. Since

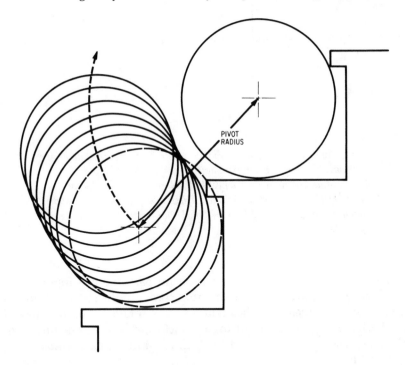

Fig. 7-29 *For lipped stairs, wheel may rest on underside of lip, unable to reach riser (dotted line). Arc of rise of rear wheel then clears lip. Note: This happens here for square lip and wheel radius scaled as shown. In general, this clearance can be assured by increasing pivot radius. Pivot radius can be increased by increasing horizontal separation of centers; vertical separation is fixed at step rise dimension.*

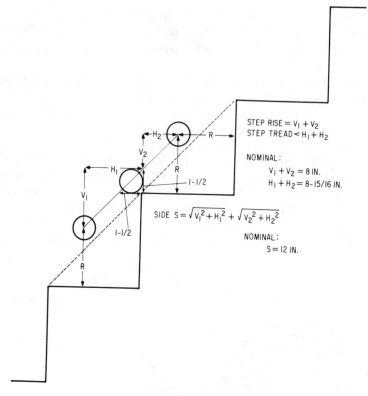

STEP RISE = $V_1 + V_2$
STEP TREAD < $H_1 + H_2$

NOMINAL:

$$V_1 + V_2 = 8 \text{ IN.}$$
$$H_1 + H_2 = 8\text{-}15/16 \text{ IN.}$$

SIDE $S = \sqrt{V_1^2 + H_1^2} + \sqrt{V_2^2 + H_2^2}$

NOMINAL:

$$S = 12 \text{ IN.}$$

Fig. 7-30 *Solving for wheel radius required to ensure stair-crest clearance for simple triangle mounting frame capable of shrouding 1-in. radius pulley at each wheel hub. Note: Solution calls for 10-in. diameter (5-in. radius) wheels. These also provide an excellent transition from rotational to pivotal modes for torque-producing drive elements.*

it must continue forward, it guides itself (in part due to slipping from imperfect traction) precisely into the front corner of the step. This forces the rear wheel (which, too, has had all of its rotational motion linked into the leapfrogging rotation) to slip backwards a fraction under an inch. This is adequate to extend the leapfrogging arc past contact with the lip of the stair.

Even though this gives a suitable solution to the geometry of the triangle of wheel centers, a simple triangle can still get us into trouble with the lip of a stair if the wheel radius is too small; in this case, the triangle can hit the crest of the stair. The problem is complicated by a requirement for a timing belt pulley somewhere along a line extending from each apex to a point halfway along each opposite side.

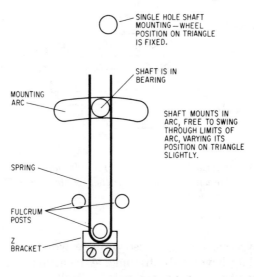

SINGLE HOLE SHAFT
MOUNTING—WHEEL
POSITION ON TRIANGLE
IS FIXED.

SHAFT IS IN
BEARING

MOUNTING
ARC

SHAFT MOUNTS IN
ARC, FREE TO SWING
THROUGH LIMITS OF
ARC, VARYING ITS
POSITION ON TRIANGLE
SLIGHTLY.

SPRING

FULCRUM
POSTS

Z
BRACKET

Fig. 7-31 *Arc mounting option for wheel hub on triangular wheel drive.*

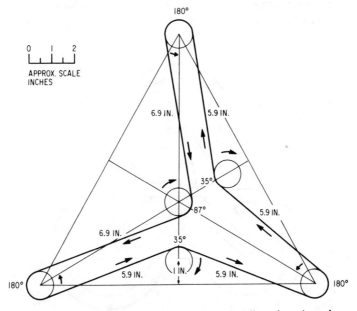

Fig. 7-32 *Final geometry of drive belt and pulleys for triangular wheel mechanism. Note: Rotation of center (driver) pulley is opposite that of wheels, pivot, and other pulleys on wheels but the same as other two idler pulleys.*

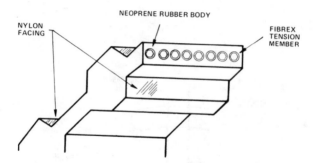

NYLON
FACING

NEOPRENE RUBBER BODY

FIBREX
TENSION
MEMBER

Fig. 7-33 *Uniroyal Twin Power® HTD belts have teeth on both surfaces or otherwise parallel construction of standard HTD belts. (Courtesy of Uniroyal)*

In any case, we must also provide worst-case clearance for a pulley extending beyond the triangle. A measurement of 1-1/2 inches in each (horizontal, vertical) dimension permits a total pulley plus belt diameter of about 2 inches—ample for our needs.

Once again, sparing you the rigors of the extensive math, this yields a minimum wheel radius of 4.95 inches. A 10-inch wheel diameter both suits this requirement and gives us a good chance to find a commercially available part.

0 1 2 3
APPROX.
SCALE
INCHES

Fig. 7-34 *Final configuration of triangular wheel drive mechanism, looking out from the inside.*

By the way, if we mount the wheel shaft in a small arc with springs rather than in a simple hole, the slippage we required earlier might be accomplished without stress on the floor (see Fig. 7-31 on p. 63).

Details of the selected geometry are spelled out in Fig. 7-32 on p. 63. Once again, Uniroyal Twin Power® HTD timing belt drives provide the necessary components. This time, though, we can use much smaller components.

The belt (see Fig. 7-33) is Model TP450L050: 3/8-inch pitch, 1/2 inch wide, 45 inches long, teeth on both sides. All six pulleys are identical, Model 10L050: 1.194-inch pitch diameter, 1-7/16-inch flange diameter, 3/4-inch flange width, 1-1/8-inch overall width. These pulleys need no bushings and can be ordered with bores (to mate to shafts) from 3/8 inch to 9/16 inch. (See Fig. 7-34).

The belt weight is 0.22 pounds, and the pulley weight is 0.28 pounds. So we appear to have, at last, a lightweight drive mechanism capable of doing everything we've asked of it.

8
The Chassis

There's a world of problems between the drive and the driven. The world of solutions has to happen in the careful design of the mechanical interface between the two.

The approach presented here is one of isolation, absorption, and protection. Much like the mystical Chinese Box Puzzle, we will nest a box within a box with nothing between them—save bumpers, shock absorbers, and a lot of careful planning.

A caution before we embark: the examples given are just that—examples only. There is no shortcut to sitting down with the real parts you have to design with and figuring out how *not* to run yin into yan as the mechanism clangs along.

In Chap. 7, in which the main mechanical drive was described, we saw that an inverted solid trapezoidal geometry seemed to make the most sense for the subchassis. The *carriage chassis* is a pan of approximately the same shape that rests within the subchassis, mounted to it with a number of tricks of the mechanical designer's trade. The main android body—or the trunk or thorax and parts north—are then mounted on the carriage chassis.

All of the shock absorbing mechanisms are mounted between the subchassis and the carriage chassis; the two, in fact, are *not* rigidly attached to each other *anywhere*.

The problem is that the ride can be very bumpy and that the shocks of traveling over stairs or uneven ground can raise unreasonable havoc with electrical connections, images presented to the visual system, items being carried, and the fasteners holding the beastie together.

Your car can tell you a lot about the vibration problem, especially if you've ever had your shocks or springs go bad. With bad springs, every bump on the road reaches you, rattles the car, and sends jolts through your bones. With bad shocks, the car seesaws or yo-yos along, takes turns like a drunken stumblebum, and turns bad bumps into bouncing hops.

But automotive shocks and springs are too big, both in size and in load characteristics, to do much good for our android. Motorcycle

66

parts tend to be long and narrow, whereas short and squat is more along the lines of android designs.

Bicycles don't have shocks, as a rule, but there are a couple of bicycle springs worth mentioning—the coil springs, usually acorn shaped, under the seat. These are usually mounted with bolts through their centers, or with a special yoke brace plate below. Remember, springs store, then release kinetic energy. Used as simple mounting hardware, these springs offer no damping, which is what shock absorbers do. They do offer one other utility, though, if specially selected: choose springs that permit some side-to-side play as well as up-and-down play and you can use them as the nonrigid link between inner and outer chassis shells. We'll cover this a little more directly in a moment.

The point we're discovering here is that standard, easily available commercial shocks and springs don't quite suit the needs of our android. Fortunately, this is a problem that industrial and manufacturing facilities have faced before.

One of the leading manufacturers of industrial shock absorbers is Ace Controls, Inc. (23435 Industrial Park Drive, Farmington, MI 48024). Some of the explanations and illustrations in this chapter are derived from Ace literature.

The problem of smoothing the ride for our android is one of providing controlled linear deceleration (see Fig. 8-1). As an analogy, consider the ride a passenger enjoys (?) in a car heading for a brick wall; his enjoyment of the ride is largely dependent on the manner in which the driver applies the brakes.

If the driver applies the brakes too lightly, the car is still moving forward when it reaches the wall, and all the energy of motion must be dissipated in the collision of the car against the wall (ouch!). The resultant *high deceleration force at the end of travel* is quite uncomfortable. Analogously, bumpers, cushions, and springs also end the travel of colliding members abruptly—not as abruptly as would be the case in their absence, but still far from the ideal case.

If the driver hits the brakes hard (panic stops) too early, the car won't collide with the wall, but the act of braking itself throws the passengers forward suddenly. This *high deceleration force at the start of travel* is analogous to the sudden stopping forces that hydraulic or pneumatic cylinder cushions and dashpots apply.

So we'll send our driver back to school and teach him to brake smoothly and surely, just firmly enough to stop the car before it hits the wall. This *linear deceleration throughout travel* lets the passengers feel a mild, continuing forward thrust with no sudden peaks. Ace has designed a shock absorber that provides linear deceleration, which is *not* a trivial design problem.

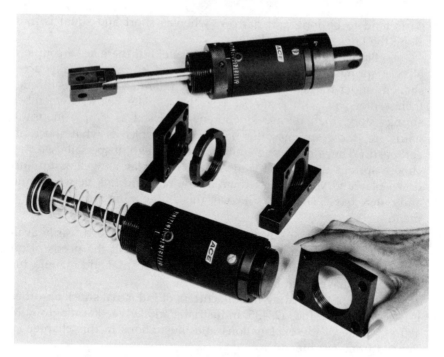

Fig. 8-1 *Ace Adjust-a-Shock shock absorbers are self-contained linear deceleration systems. The mounting flange can be threaded at either end of the main cylinder housing. The dial near the center of the shock adjusts the exact decelerating force applied to meet the demands of a wide range of loads. (Courtesy of Ace Controls, Inc.)*

Shock absorbers work by forcing a piston down a cylinder and displacing a viscous fluid, one that resists its travel. This fluid must escape through an orifice. The size of the orifice, the speed of the stroke, and the mass behind it all determine the deceleration force applied by the shock.

These deceleration forces are not linear with respect to time for any fixed orifice size; for linear deceleration, the orifice size has to change at the same rate that the velocity of the moving object changes.

While it's next to impossible for the size of a single orifice to be metered through needle valves or similar mechanisms in this fashion, there is another approach. It requires a number of orifices spaced along the length of the cylinder. As the piston travels the length of the cylinder, it sequentially closes off orifices, effectively decreasing the total orifice area.

The problem with the multiple orifice approach is that it provides true linear deceleration for only one specific combination of speed, mass, and propelling force.

NEW ACE "SLOT-ON-HOLE"
ADJUSTABLE ORIFICE DESIGN

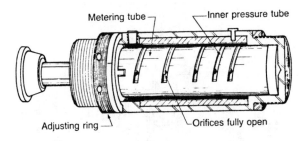

KNIFE-EDGE EFFECT
WITH "SLOT-ON-HOLE" METERING

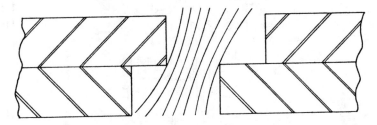

Fig. 8-2 *Orifice design of Ace shock. (Courtesy of Ace Controls, Inc.)*

Ace has solved the problem of making this group of factors variable within a single shock with the unique design of their Adjust-A-Shock® (see Fig. 8-2). A metering sleeve with slanted slots fits over the fixed inner cylinder. Rotating this sleeve variably masks the orifice, effectively increasing or reducing its size. A slot is provided over each orifice. Thus, rotating the sleeve adjusts the shock for various load requirements.

The Ace Model A 1/2 × 1 is a self-contained spring-return shock, ready to mount on its supplied flange, which can be threaded onto either end (see Figs. 8-3 and 8-4). It has a 1/2-inch bore and a 1-inch

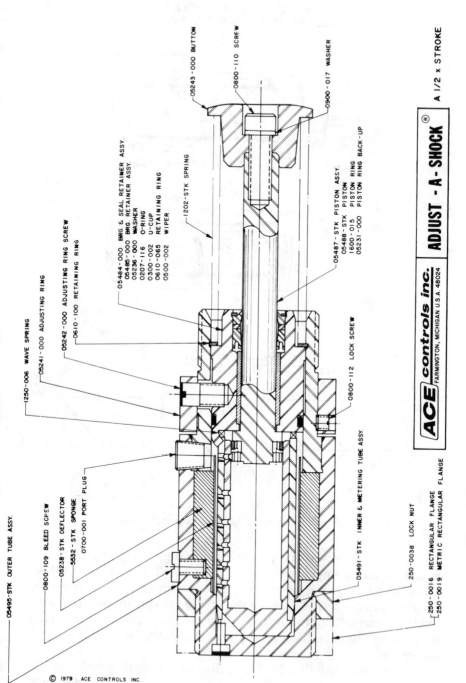

Fig. 8-3 *Cross section of Adjust-A-Shock. (Courtesy of Ace Controls, Inc.)*

05243 - 000 BUTTON

0800 - 110 SCREW

0900 - 017 WASHER

05484 - 000 BRG. & SEAL RETAINER ASSY.
05485 - 000 BRG. RETAINER ASSY.
05236 - 000 WASHER
0207 - 116 O-RING
0300 - 002 U-CUP
0610 - 065 RETAINING RING
0500 - 002 WIPER

1202 - STK SPRING

05487 - STK PISTON ASSY.
05488 - STK PISTON
1600 - 015 PISTON RING
05231 - 000 PISTON RING BACK - UP

1250 - 006 WAVE SPRING

05241 - 000 ADJUSTING RING

05242 - 000 ADJUSTING RING SCREW

0610 - 100 RETAINING RING

0800 - 112 LOCK SCREW

05495 - STK OUTER TUBE ASSY.

0800 - 109 BLEED SCREW

05238 - STK DEFLECTOR

5532 - STK SPONGE

0700 - 001 PORT PLUG

05491 - STK INNER & METERING TUBE ASSY.

250 - 0038 LOCK NUT

250 - 0016 RECTANGULAR FLANGE
250 - 0019 METRIC RECTANGULAR FLANGE

ACE controls inc.
FARMINGTON, MICHIGAN U.S.A. 48024

ADJUST - A - SHOCK ®

A 1/2 x STROKE

© 1979 ACE CONTROLS INC.

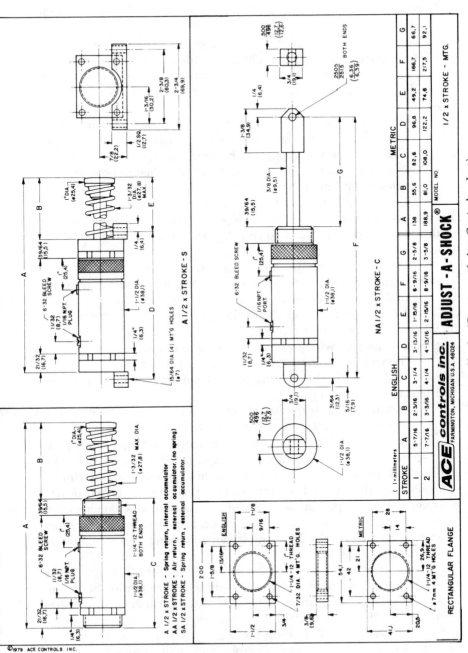

Fig. 8-4 Details of Adjust-A-Shock. (*Courtesy of Ace Controls, Inc.*)

stroke. It can handle up to 1,000 inch-pounds per cycle, up to 750,000 inch-pounds per hour—more than enough for most circumstances an android will encounter.

We will be using four of these shocks. There's one more small shock we have to consider, though: these little beauties cost about $50 apiece. If you're hesitant about the investment, investigate away. In the end, you may agree with me that their advantages outway their cost.

The Adjust-A-Shock, as you can see from Figs. 8-1 through 8-4, ends in a button. In our design, the shock is flange-mounted to the carriage chassis (upper) shell, and the buttons rest on a reinforcing plate in the main mechanical drive subchassis (lower) shell, where there is no direct mechanical mounting.

Just as a failsafe, and to avoid damaging the shocks in case of a severe impact, rubber bumpers or blocks at the full-compression end of travel clearance keep the two chassis shells from banging into each other.

So far, then, the two shells float freely, one inside the other. This may not be a problem in some applications, but when we consider the case of the android descending the stairs with some load in its arms, it's easy to envision the inner shell falling out or helplessly dangling by its wires. Obviously, we need something to hold the halves together. And here's one area where we can stay simple or get elegant.

The simplest solution would be to provide steel cables, bolted into one of the shells at each end, to limit the amount on play at any selected number of points.

Another option is the use of extension springs, mounted in the same manner as the cables. Alternatively, there is another spring configuration called a *drawbar*. Here, wires are looped through a compression spring and hooked around the end opposite the end each wire enters. This allows spring force to be applied through to a maximum extension and then provides a firm stop. If the ends are bolted down, only vertical play is permitted. If the ends are mounted to bolted-down rings, an additional degree of freedom or play results. Still another degree of freedom results if the end is looped around a rod or stiff wire length secured at its ends.

Another option involves circles of rubber—these might be cut from inner tubes, tires, or rubber balls—drilled for mounting hardware at opposite ends. The elasticity of the rubber (or whatever synthetic has been used in its place) permits a number of degrees of freedom plus sufficient tensile strength to keep the shelves from parting. Depending on the thickness and strength of the rubber piece you provide, you may choose to use this tension-loop approach in several places, or in one central position.

As you mull over your options, consider readily available surplus mounting components. One source for these gems is Jerryco, Inc. (5700 Northwest Highway, Chicago, IL 60646). Their catalog of "Nugatory Contrivances" is a gadget lover's dream book.

While you're picking up catalogs, there are a number of commercial sources of drive and mounting components. Most will supply catalogs on request; others have nominal charges: Stock Drive Products, Division of Designatronics, Inc. (55 South Denton Avenue, New Hyde Park, New York, NY 11040); Pic Design, a Benrus Division (P.O. Box 335, Benrus Center, Ridgefield, CT 06877, and 6842 Van Nuys Boulevard, Van Nuys, CA 91405); Winfred M. Berg, Inc. (499 Ocean Avenue, East Rockaway, Long Island, NY 11518); and Boston Gear Division, Rockwell International (14 Hayward Street, Quincy, MA 02171) are among the leading sources.

While we're on the subject of mechanics, you might want to consider some nonhardware fasteners as an alternative. These can—if properly selected and applied—offer tremendous savings in weight and construction time.

The cyanoacrylate glues that can stick football players' helmets to goalposts, send trapeze artists swinging on broken boards, and lift boats with blocks have been improved in both formulation and packaging over the years. One example is the Bondo® Instant Glue Pen. This is a drop-at-a-time applicator using a specially designed spring-loaded tip. It's great for bonds between metal and metal or rubber, many plastics, and lots of other android-building tasks.

A tip for the cautious: While it still bonds skin as easily as ever, simple nail polish remover quickly cleans it away. And a tip for the aggressive: even surfaces that don't normally bond well with cyanoacrylates can be mated if they are first thoroughly scrubbed with alcohol.

For porous surfaces and places where an epoxy is suggested, new twin-syringe packaging helps meter an exact proportion of the two components automatically with the push of a plunger.

Contact cement is a quick and simple way of mounting items that aren't likely to be subjected to shear or stress other than compression. It might be useful, for example, for mounting the bumpers to the subchassis, for mounting external bumpers around the subchassis, or for a ribbon switch to act as a collision sensor (more on this later).

Another interesting substance, though not an adhesive in the sense of the earlier examples, is plastic rubber. This is excellent for waterproofing and sealing joints in metal, caulking, sealing wires in through-holes, some insulating, and in providing a rim-bead to protect your fingers from the jagged edges of cut metal.

Now what of the carriage chassis itself? So far we have concentrated on its relationship to the main mechanical drive subchassis. It's time to consider it as its own entity. Remember, the carriage chassis is the isolating link between the drive chassis and the android's upper body. Let's consider for a moment what the mounting of the upper body presents for us in terms of problems and possibly opportunities.

The trunk or thorax of the android may be any shape at all, but a number of considerations suggest cylindrical. That's because any shape will have to rotate, so rotational clearances have to allow for the maximum radial dimension. The maximum volume within the trunk is available for electronics if the maximum rotational radius is maintained at every angle. This describes the circular cross section of a cylinder.

There's a very inexpensive way to allow for rotation of the cylinder—"lazy susan" ball bearing turntables. These are available at this writing for under $5.00 from Edmund Scientific (101 East Gloucester Pike, Barrington, NJ 08007). These are available with load ratings to 1,000 pounds, more than enough for our android, yet are lightweight and dimensionally compact.

There still remains the problem of allowing the rotating trunk the capability of leaning forward and backward as much as 45°. We can accomplish this by mounting the bottom of the rotating turntable to an axle-mounted brace of some sort. An excellent choice might be the flywheel from a small car, cut parallel to its diameter an inch or so from center. This not only provides the right kind of pivoting mount, it also allows the matching starter gear to be used as the pivot driver's reduction gear (because, of course, the pivot motor output has to be low rpm and high torque).

Other similar pivot-on-axle mounts also could be used; and other drive mechanisms, like toothed belt and pulley drives. The important considerations are control, backlash, and braking. Because of the high reduction ratio between the motor speed and the actual end tilting (pivot) speed, simple motor braking through shorting of the motor leads (assuming a permanent magnet dc motor, which can be braked by using a relay to short its leads together when it isn't being driven) may prove adequate.

Within the carriage chassis, we must be certain to allow adequate clearance for the pivoting trunk at every angle of tilt. Note in Fig. 8-5 that the battery indicated, which is typical of the battery size the android is likely to require, must be mounted at a slight angle to permit full clearance.

Once you have arrived at the final dimensions for your own beastie, you face a fastidious technical task in determining the exact cutting and bending dimensions for the sheet metal or material you select.

As with the other aspects of material selection, we face important tradeoffs in strength, weight, availability, and ease of handling.

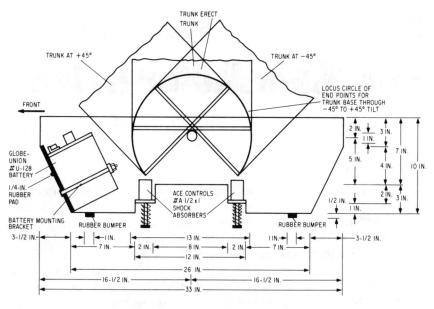

Fig. 8-5 *Side view of carriage chassis.*

Even once you think you've found the right materials, talk to the experts—in this case, a machine shop might be the best place to find them—about variations and tricks of the trade that might help you out.

Aluminum, for example, is not one metal but a family of alloys. Even once you've found the right aluminum alloy, tricks like the right way and the right places to bend it, reinforce it, and so on can mean the difference between success and failure.

It's hard to know which is the cart and which the horse when designing something as eventually complex as an android. You almost have to know what goes inside the chassis before you can exactly determine its dimensions and characteristics, while its dimensions and characteristics are a deciding factor on your selection of where to mount what and what shapes and sizes various components must not exceed.

My best advice is that your primary tools in designing and building the chassis—and the rest of the android, for that matter—are a pencil, a bunch of paper, and—yes—an eraser.

9
The Main Motor Drive

Decisions about a suitable drive motor and appropriate surrounding mechanisms, like speed and direction control, linkage to the driven elements, etc., have to be made early in the design process, even before we have many of the final answers to questions crucial to the motor selection itself, like the final weight of the unit.

So we'll start with what we do know, find what we need to know, and see what we can get away with otherwise. One thing we can easily determine is how fast we want our mechanism to go at top speed; from that, we can easily determine the rotational speed of the final drive.

We are going to deal here in linear speed in terms of feet per second. To convert from miles per hour:

$$\frac{\text{feet}}{\text{second}} = \frac{\text{miles}}{\text{hour}} \times \frac{1 \text{ hour}}{60 \text{ minutes}} \times \frac{1 \text{ minute}}{60 \text{ seconds}} \times \frac{5,280 \text{ feet}}{1 \text{ mile}}$$

$$= \frac{\text{miles}}{\text{hour}} \times \frac{5,280}{3,600}$$

$$= \text{mph} \times 1\,{}^{7}\!/_{15}$$

$$= \text{mph} \times 1.467$$

It took some legwork with a stopwatch in hand to determine how specific speeds relate to human routines, such as walking and running. Here are the results, which you may want to ascertain yourself:

Slow walk   3 ft/s (feet per second)
Stroll   4.5 ft/s
Briskly paced walk   6 ft/s
10-minute mile   8.8 ft/s
Run   10 ft/s
Full sprint   15 ft/s

To translate linear speed to rotational speed, we'll need to know the diameter of the final drive wheel or pulley. For the track drive we discussed, the pitch diameter of the drive pulley is 5.73 inches. For the

76

Table 9-1 *Rotational speeds versus linear speeds for three drive diameters.*

Linear speed (feet/second)	5.73-inch pulley (multiplier 40.0)	10-inch wheel (multiplier 22.9)	23.9-inch pivot (multiplier 9.6)
1.0	40.0 rpm	22.9 rpm	9.6 rpm
2.0	80.0 rpm	45.8 rpm	19.2 rpm
3.0	120.0 rpm	68.7 rpm	28.8 rpm
4.0	160.0 rpm	91.6 rpm	38.4 rpm
4.5	180.0 rpm	103.1 rpm	43.2 rpm
5.0	200.0 rpm	114.5 rpm	48.0 rpm
6.0	240.0 rpm	137.4 rpm	57.6 rpm
7.0	280.0 rpm	160.3 rpm	67.2 rpm
8.0	320.0 rpm	183.2 rpm	76.8 rpm
8.8	352.0 rpm	201.5 rpm	84.5 rpm
9.0	360.0 rpm	206.1 rpm	86.4 rpm
10.0	400.0 rpm	229.0 rpm	96.0 rpm
11.0	440.0 rpm	251.9 rpm	105.6 rpm
12.0	480.0 rpm	274.8 rpm	115.2 rpm
13.0	520.0 rpm	297.7 rpm	124.8 rpm
14.0	560.0 rpm	320.6 rpm	134.4 rpm
15.0	600.0 rpm	343.5 rpm	144.0 rpm

triangular wheel drive, we have two diameters to consider. The first of these is the 10-inch diameter of the wheels; the second is 23.9 inches. (The second diameter is determined as twice the sum of the distance from the triangle center to the wheel center and the wheel radius; in other words, the distance from the triangle center to the far edge of a wheel is the radius for the pivotal motion.)

Recognizing that rotational speed and linear speed link through the circumference of the drive wheel or pulley, we can mathematically relate rotational speed to linear speed and diameter:

$$\frac{\text{Revolutions}}{\text{minute}} = \frac{\text{feet}}{\text{second}} \times \frac{60 \text{ seconds}}{1 \text{ minute}} \times \frac{1 \text{ revolution}}{\pi \times \text{D inches}} \times \frac{12 \text{ inches}}{1 \text{ foot}}$$

$$= \frac{\text{feet}}{\text{second}} \times \frac{60 \times 12}{\pi \times \text{D}}$$

$$= \frac{\text{feet}}{\text{second}} \times 229.2 \times \frac{1}{\text{D}}$$

Calculating the multiplier for each of these two drive types, we find the 5.73-inch pulley drive multiplier is 40.0; for the 10-inch wheel drive it's 22.9, and for the 23.9-inch pivot it's 9.6. Table 9-1 extends these multipliers over an extensive range of linear speeds (in terms of what we'll actually end up using), giving the drive rpm for each instance.

Notice, by the way, that the pivoting action of the triangular wheel drive translates rotational speed into a faster linear speed than that produced when driving along a flat surface. As a result, it climbs faster than it rams. Unfortunately, it also mandates higher torque, which we'll investigate shortly.

First, we're going to share some of the wisdom that motor manufacturers have passed along. The following quotes are reproduced with the permission of each company; additionally, copies of the specific pieces of manufacturer's literature—or updated versions—are generally available on written request.

One of the leading manufacturers of reasonably priced fractional horsepower motors and gearmotors is the R.A. Boehm Company, subsidiary of Baldor (another leading name), Fort Smith, AR 72902. In their Data Section & List Price Sheet, "Fractional & Subfractional Horsepower Motors and Gearmotors," the following excellent description* of the significance of torque appears:

LET'S TALK TORQUE
(or . . . Breaking the Horsepower Habit)

When applying and specifying larger fractional and integral horsepower motors, it is common to think and specify only horsepower and speed. However, when applying sub-fractional horsepower gearmotors, the speed and torque required at the output shaft is the proper data for selection of the right unit.

TORQUE IS WHAT MAKES IT GO AROUND

Regardless of the speed of operation, it takes torque to turn the machine. On most machines, the torque required is independent of speed.

The first consideration: How much *torque* is required?

The next should be: What is the maximum *speed* at which the machine will run and how fast will the gearmotor have to turn to achieve this speed?

Those two questions form the real basis for gearmotor selection.

---

* Copyright © 1979 by Baldor. Reprinted by permission.

OTHER APPLICATION CONSIDERATIONS

Beyond these two basic items, additional details such as hours of operation, conditions around the motor, voltage and mechanical configuration to suit the machine complete the information required to select the unit best suited for your application.

Torque is a twisting force applied over a distance, whether it be applied to a lever, a wheel, or anything free to rotate about an axis or fulcrum point. The amount of force and the amount of distance determine the amount of torque:

Torque = force × distance

The normal units of torque are inch-ounces, inch-pounds (1 inch-pound = 16 inch-ounces) and foot-pounds (1 foot-pound = 12 inch-pounds = 192 inch-ounces). Weight is a primary consideration in determining how much torque is required of a motor, but other forces also are present: these include frictional forces, drag, and so forth. Most important, we must consider the acceleration we will require from our motor. The force required to accelerate a mass at a given rate is determined by the favorite formula of high school physics:

Force = mass × acceleration

Now we're getting somewhere! (You'll catch on in a second.) Notice that in the previous expression for torque, torque does not vary with speed. From the expression for force, we can see that it does vary with acceleration. (In fact, even with slow, constant acceleration we can eventually reach any speed. The motor eventually reaches its no-load speed—or nearly does—as the mechanism accumulates momentum.)

There are two conditions under which we are especially concerned with the value of this acceleration. One is in determining how quickly our android has to reach its top speed (or possibly a greater value of acceleration may be necessary for some maximal change of speed in some minimal length of time). The other—of greater concern right now—is the acceleration required to counter the acceleration due to gravity, 32.2 feet/second². In other words, before we calculate torque, we'll have to calculate the forces involved.

The worst case (one which we will neither actually encounter nor design for) would involve moving straight up against gravity, where the weight of the mechanism takes its worst toll. To merely hold its own against gravity, the necessary force is determined by the mass in slugs times gravitational acceleration. (*Pounds* are units of weight, which is a measure of *force*: the *poundal* is a derived unit of force, the force that accelerates the mass of a 1-pound object by 1 foot/second². More commonly, the *slug* is a derived unit of mass, defined as the mass that 1 *pound* of force accelerates to 1 foot/second². Are you confused?) Actually, it means a force equal to the weight of the mechanism.

The actual case is determined by finding the gravitational component of acceleration, which is the weight times the sine of the ramp angle. For the 45° ramp of a stairway, this is 0.707 times the weight, or 70.7 pounds per 100 pounds of android weight.

The second accelerating force is that required to cause forward motion. How soon do we want to go how fast? For example, to reach a velocity of 15 feet/second in 1 second, the acceleration is 15 feet/second². To reach 10 feet/second in 20 seconds, the acceleration is ½ foot/second².

I'm going to skip a few intermediate steps and simply state that I've found a reasonable acceleration requirement to be between 1 and 2 feet/second². If you can share a peek at my crystal ball, you'll see that faster acceleration means a heftier motor, a bigger battery, heavier duty drives, and so on.

To calculate the force we need for this acceleration, we don't have to use *poundals* or *slugs*; instead, we can adjust for pound units by canceling the gravitational factor in a simple division:

$$\frac{\text{Weight}}{\text{Gravity}}$$

Now, the force needed to accelerate without climbing is:

$$\text{Force} = \frac{\text{weight}}{\text{gravity}} \times \text{acceleration}$$

This calculates out to 3.1 pounds per 100 pounds of android weight at an acceleration of 1 foot/second²; or 6.2 pounds per 100 pounds of android weight at 2 feet/second².

Summing this with the force necessary to fight gravity in a climb, we find that we require a total force of 73.8 to 76.9 pounds per 100 pounds of android weight.

We can continue in much the same fashion and calculate the torque we'll need (force × distance). The distance(s) involved is the radius of the drive wheel or pulley. This is 2.87 inches for the track drive pulley, 5 inches for the triangular wheel drive when wheeling, 11.9 inches when pivoting.

When you've gone through the calculations, you'll find a range of torques from 211.8 to 915.1 inch-pounds per 100 pounds of android weight. This, in turn, projects out to a range from 211.8 to 1,830.2 inch-pounds for a range of weights from 100 to 200 pounds.

That's quite a range! But hang in there and we'll nail it down in just a bit. Unfortunately, the range is going to have to increase first as we calculate the motor horsepower we'll need.

Our friends at Boehm/Baldor have simplified the relationship of torque, speed, and horsepower to:

$$\text{Horsepower} = \frac{\text{speed} \times \text{torque}}{63,000}$$

Taking a range of average speeds between 5 and 10 feet/second, the range of our horsepower requirement is between 2/3 and 2.8 horsepower. (This range is approximate, but then so is the rest of this.) Distributed over four motors, this implies motor horsepowers of 1/6 to 3/4 horsepower. If you think the horsepower range is large, I suggest you try pricing these motors!

Okay, let's try to set some design goals to help sort this problem out into something even more reasonable.

We've seen that torque and horsepower requirements climb with weight, acceleration, and speed. So let's peg our design-center weight at 150 pounds, acceleration at 1 foot/second^2, top speed at 8 feet/second, and average speed at 4 to 4.5 feet/second.

We also can split our investigation into separate calculations for motor requirements for the tracked system and for the triangular wheel drive.

First, for the tracked mechanism, a top linear speed of 8 feet/second means a rotational top speed of 320 rpm. The climbing force and motive force total 110.7 pounds. The necessary pulley torque is 317.7 inch-pounds total, or 79.4 inch-pounds at each of four motor-pulley combinations. For convenience, we can recognize this as equal to 6.6 foot-pounds or 1,271 inch-ounces. The horsepower requirement is 0.8 to 0.9 horsepower, or 1/5 to 1/4 horsepower per motor. Not bad.

For the triangular wheel drive, we discover an interesting phenomenon. The top rotational speed is 183.2 rpm in the wheeling mode, 76.8 rpm in the pivoting mode. The force required is *again* 110.7 pounds, in *both* modes. The torque requirement in the wheeling mode is 553.5 inch-pounds (or 46.1 foot-pounds or 8,856 inch-ounces); in the pivoting mode it's 1,317.3 inch-pounds (or 110 foot-pounds or 21,077 inch-ounces). The horsepower requirement is *again* 0.8 to 0.9 horsepower, or 1/5 to 1/4 horsepower per motor in each of four motors.

Aha! Horsepower follows weight! And the same horsepower can yield a greater torque at a lower speed. But alas, we are going to have to either compromise top speed or motor size to both roll and pivot with our triangular wheel drive. Or we may have to put our beastie on a diet.

The reason is that the radius against which torque is calculated for the rolling mode is only 5 inches, while it's 11.9 inches for the pivoting mode. And since the pivoting rotational speed is the same as the wheeling rotational speed, when we require an 8 feet/second top speed in the wheeling mode, it means a rotational speed of 183.2 rpm, which corresponds to a 19.1 feet/second top speed in the pivoting mode. Similarly, the 4 to 4.5 feet/second average speed we're designing for becomes 9.6 to 10.8 feet/second. And this brings our horsepower requirement up to 1.9–2.1, or roughly 2 horsepower—meaning four 1/2-horsepower motors, no doubt.

But look out! If you want to have your cake and eat it too, your tab can easily top $500 . . . or more. A suitable motor, for example, might be the type 701-HAT from the Dumore Company (1300 Seventeenth Street, Racine, WI 53403). These motors offer 50–210 inch-pounds of torque (we need 330) and speeds from 15 to 800 rpm (we need 185). The weight of each of these motors is 12¼ pounds, or 49 pounds for four. With motors this heavy, it's a good idea to allow for some "bonus" torque, so a 3:1 to 4:1 ratio drive to the hub (through timing belts and ratio-diameter pulleys) is a good idea, giving us the speed and torque we need.

The bad news is a 100-unit minimum order requirement; in light of it, quoting a price for the motor seems almost futile. Still, for those of you in a position to spend anything (or somehow get anything), this is one way to get both speed and torque.

Keep in mind, though, that a more powerful motor needs a heavier battery, a heavier duty drive, heavier connecting cables, and so on.

But don't give up yet!

There's a delightful operation in Philadelphia, famous far and wide for good values on unusual electronic and mechanical finds. And I found workable motors there for under $10.

Herbach & Rademan, Inc. (401 East Erie Avenue, Philadelphia, PA 19134) publishes a lovely catalog called *This Month*. Well, one month, an item caught my attention:

High Torque Reversible 6 or 12 Volt PM Field Motor

Originally designed as a "pedal assist" drive motor for bicycles . . . This beautiful, chrome finished, high efficiency, reversible PM field DC motor will find many applications in mechanical drives (wherever a 6 or 12 VDC battery or supply can be available) such as an assist for wheeled vehicles, driving mechanisms and other mechanical duties where high starting torque is needed. Can also serve as a low power DC generator utilizing a blocking diode to prevent battery discharge through motor. *No load speed approximately 3600 RPM, current 3.5 Amps. Will supply considerable torque (at reduced speed) with 6 VDC. Stall torque is 15 in/lbs at 12 VDC, 50 Amps. Rated .194 HP at 3150 RPM (approximately).* Use motor with gear or chain drive to increase torque at reduced speeds. Battery should be capable of momentary current loads in excess of 50 Amps for maximum torque output. Motor will operate for extended periods without overheating. (Duty cycle dependent on amount of loading.) Oil-less sleeve bearings. Shaft, 5/16" dia. × 1-3/8" with 1/2" wide × 1/2" long flat. Size: 2-1/2" dia. × 5-1/8" excluding shaft. End mounts on #10 threaded studs. Wire lead connections. Reversible with D.P.D.T. cross-wired switch. New. Shipping Weight each, 5 lbs.

When I bought mine, they were $9.95 each, 6 for $54.00, 12 for $96.00. Better double-check those prices if you're interested now, though. Herbach & Rademan sell surplus only, so they may be out of the ones I bought but have a substitute handy.

By now, you should be wondering how an under-1/5-horse-power motor can do the 1/2-horsepower job we've been investigating. The answer is—it can't. But it can do most of the job, if we're willing to make a few compromises.

In terms we're more used to hearing from doctors, we're going to have to slow down and try losing a few pounds.

Actually, that's not too difficult a line of thinking to agree with. Consider speed, for example. We've already noted that 6 feet/second is a fairly briskly paced walk. Now say we peg our top speed for the android at something like 6½ feet/second (instead of 8 feet/second, which we previously had used as a working figure. This corresponds to a wheel speed of about 150 rpm. Compared to the 3,600 rpm maximum speed (the no-load speed is the maximum speed the motor can eventually attain, once momentum is working for it and inertia is dead, in un-engineering lingo), this implies a 24:1 reduction. In the real world, a 25:1 reduction might be easier to reach; this still gives us 6.3 feet/second for a top speed.

Philosophically, it's probably just as well that the android will be able to keep up with any walker, but anyone will be able to outrun it.

The 25:1 reduction in speed rewards us with a 25:1 increase in torque (theoretically—in practice reducers tend to eat some of that up). The stall torque, which is the maximum output torque, climbs from 15 inch-pounds to 375 inch-pounds, a total of 1,500 inch-pounds for four motors. So if we felt safe climbing a stairway at stall loads, we could haul our 1,320-inch-pound torque load upstairs at will.

Alas, the 200-ampere load would kill most batteries by the time the android is halfway up the stairs.

But if we can trim our target weight down to about 120 pounds (from the original 150), we can trim the torque requirement to under 1,100 inch-pounds. This brings the current requirement down to about 12 amperes per motor, or 48 overall. Not great, but much more probably acceptable. It probably means the android will have to huff and puff its way upstairs and plug in for a recharge almost immediately.

The most important lesson in all of this is probably that nothing saves when building androids like saving weight.

10
The Battery

You may decide that an android should be very much like a human. After all, there are a number of schemes available for converting heat into electricity. And if our bodies can burn food for fuel, why not give the android the same capability?

It took nature a long, long time to develop digestion for us poor humans—and even now, we know it must occasionally malfunction, or there would be no such thing as antacids. Someday, direct consumption of tangible food-fuel may indeed be the best way to keep an android going. But for now, I'm not quite willing to discount nature's millennia-long head start.

Others may think of solar power. First, I would like to refer you to my "Solar Powered Night Light," published in "Three Unique Projects" (*Radio-Electronics*, April 1979). Second, I would like to invite you to investigate the surface area versus output power characteristics of the best (which are, as might be expected, also the most expensive) of today's commercially available solar cells.

Not that solar power is unfeasible—nor is fuel, per se. For example, a solar-powered station could electrolyze water into its component hydrogen and oxygen gases. (In fact, solar power also could be used to boil water in a simple still to provide pure distilled water for electrolysis.) Solar heated steam generators or motors could drive a pump to compress and bottle these gases. The android could then use the bottled gases in an on-board fuel cell. If you wish to pursue this option, consult NASA. Fuel cells provided electricity aboard a number of space vehicles. The disadvantage to this approach is that the android must then always be within range of a fuel station when its fuel supply gets low.

An eminently more practical approach is to use on-board rechargeable batteries, an on-board battery charger, and to plug into the power mains whenever the battery "gets hungry." Proponents of solar and fuel cell power can appreciate that either approach can be used at distinct power stations (through the use of commercially available power inverters) without requiring the android to return exclusively to them.

Okay, we've pretty well set the ground rules for the on-board power system. It's going to be a battery, it's going to have its own charger on board, and there's going to be some mechanism to recognize when the battery charge is low enough to send the android in search of an outlet.

Now it's time to identify and compare the various rechargeable battery technologies and determine what our best choices are, keeping in mind our dictum that the android must not present any danger to life or property.

We will investigate five battery technologies: standard lead-acid automotive batteries, maintenance-free automotive batteries, nickel-cadmium batteries, secondary (rechargeable) alkaline batteries, and gelled-electrolyte batteries. A new (sixth) technology is emerging, secondary lithium cells, but at this writing has not advanced enough to allow us to evaluate it for android applications.

Traditional lead-acid automotive batteries are easily available, relatively inexpensive, and usable over a broad temperature range. At 2 volts per cell—higher than many other technologies offer—it makes high voltages available with fewer cells. It has no thermal runaway problems, no memory problem, no cell reversal problem as a rule, and it has an excellent reputation as a reliable workhorse.

But it needs maintenance. Its electrolyte (water) needs periodic replenishing. When overcharged, it gasses explosive and corrosive hydrogen and oxygen. It must be operated upright. And it weighs quite a bit.

The newest turn for automotive battery technology has been the "maintenance free" lead-acid battery. This battery was designed to overcome the requirement of standard automotive lead-acid batteries for replenishment of water by designing the maintenance-free batteries with excess water already contained within them. This amounts to roughly 9 ampere-hours of excess water per ampere-hour of capacity.

In these batteries, the mechanism usually responsible for depletion of water is overcharging. Fixed-voltage (or float) charging over extended periods *is* an overcharge condition for these batteries. Eventually, even the generous excess of water can be depleted, causing dry-out failure of the battery.

While manufacturers claim that the battery can be operated in any position, this is qualified by a requirement that the battery not be recharged in an inverted position.

These batteries are more expensive than standard automotive batteries, they weigh more, and they present a higher internal impedance. The worst part is that they can and do emit acid spray and gas and should not be operated near electronic circuitry or corrosion-prone materials.

Nickel-cadmium cells are maintenance free, requiring no water or chemistry to be added or replenished. They emit virtually no gas, can be operated in virtually any position, are widely available, and represent a stable, established technology.

But the operating temperature range of these cells is limited, with the available voltage per cell dropping with temperature. Cell voltage is only 1.2 volts. Charging can be a difficult proposition. The state of charge cannot easily be monitored. NiCads are fairly expensive, compared to many other battery technologies. A *memory* effect also plagues NiCads: if a cell is repeatedly discharged to a given level, later attempts to discharge past that level may fail.

Rechargeable alkaline batteries, like NiCads, are maintenance free, don't gas, and can be operated in any position.

However, they only deliver about 1/4 of the service per weight (in ampere-hours) of primary cells (standard, nonrechargeable). They can only be charge-cycled about 50 times. They have high internal impedance, and they're difficult to recharge successfully.

The last class of battery we want to look at—and I promise I've saved the best for last—is the completely sealed, maintenance-free, gelled-electrolyte lead-acid battery. These can be used and recharged in any position, operate over a very wide temperature range, exhibit no memory effect, can be charge-cycled hundreds of times, and offer excellent service per weight. They represent a stable and improving technology and are widely used in such applications as alarm systems and those unforgettable two-headlights-on-a-box-with-a-battery power failure emergency lights you keep seeing in public buildings.

Did I say *completely* sealed? They usually are. But in the case of severe heating or overcharging, like any battery, the water within them tends to electrolyze into its components—hydrogen and oxygen. Specially designed safety vents on these batteries permit these gases to escape, usually in quantities not worth worrying about. But, if the batteries are operated in a sealed enclosure that is not vented to the outside world, these gases can collect, with possibly explosive consequences!

The several android design versions we have discussed so far all require that the trunk of the android be capable of leaning frontward or backward as much as 45°. Since custom rubber bellow boots are expensive to buy and difficult to make, it seems inevitable that there will be some outside atmospheric access through the hole on either side of the trunk pivot. If this seems inadequate or of dubious merit, it's easy enough to hook a small fan up to the battery charger circuit and turn it on whenever the battery is charging; this also would help sink heat from the motor drive transistors. It's probably a good idea to select a fan motor that won't expose the vented air to commutator brush sparking.

Fig. 10-1 *Globe-Union Model U-128 Gel/Cell®. (Courtesy of Globe-Union)*

There is a great deal of very useful information about gelled-electrolyte sealed rechargeable batteries, which is available from their manufacturers. While many of the highlights will be presented here, you are encouraged to contact them.

Globe-Union Battery Division offers Gel/Cell® batteries. Write them at 5757 North Green Bay Avenue, Milwaukee, WI 53201, for their catalogs, data sheets, and *Charging Manual*.

The battery in my android is a Globe U-128, shown in Fig. 10-1. It offers 20 hours of service at 1½ amperes to 5.7 minutes at 100 amperes to about 3 minutes and 20 seconds of service at 150 amperes (see Fig. 10-2). It is designed for repetitive charge/discharge cycling, is completely self-contained, is sealed (except for pressure vents), and can even be operated upside down for extended periods—though this isn't recommended.

Gates Energy Products offers a *Battery Application Manual* with a cover price of $1.50. Write them at 1050 South Broadway, Denver, CO 80217. This 48-page reference manual, while thoroughly a "puff" piece for Gates cells, is a treasurehouse of information on charging and discharging characteristics (including graphs and examples), charger design (including a number of schematics), service life, a glossary, and more.

Those of you who read literature from both Globe-Union and Gates Energy Products will be able to enjoy a chuckle at the subtle ways these two giants try hard to endorse their own specific construction techniques at the expense of their competitors. A word to the wary: My friend Chuck Small, Instrumentation Editor of *Electronic Engineering Times*, tells me that regardless of what manufacturers say, all such cells seep out some kind of ooze (yuk!). Chuck is putting together an unin-terruptible power supply for his home computer with a few motorcycle batteries and a Rubbermaid® drainboard. His batteries are *almost* sealed: they have a little hose fitting that directs any emerging fluids or gases down a plastic tube.

The open-circuit voltage of a gelled electrolyte battery is an ex-cellent indicator of its charge status. The nominal cell voltage is about 2.12 VDC. This will be higher just after the battery is taken off charge, decreasing to about 1.75 volts as the charge is spent.

These batteries are usually rated by their 20-hour rates, which is the amount of current they can continuously deliver for 20 hours between full charge and discharge. The U-128, for example, is rated at 28 ampere-hours (1.4 amperes for 20 hours). It is not possible to inter-polate this figure to determine a linear function. The same battery can deliver 2.3 amperes for only 10 hours (23 ampere-hours), 4.2 amperes for 5 hours (21 ampere-hours), 15 for 1 hour (15 ampere-hours), and so on. The data sheet in Fig. 10-2 spells out and charts some of these key rating-points. Note that the specified discharged-condition voltage is specified between 9.6 and 10.5 VDC for this 6-cell battery (12 VDC nominal, or 12.72, depending on how picky a specifier you are), or 1.60 to 1.75 VDC per cell. The maximum current this battery can deliver (in a brief pulse under near-short conditions) is 500 amperes. At freezing, when only about 85 percent of its capacity is available, it can still deliver as much as 200 amperes for from 5 to 45 seconds.

These criteria are very important. They recommend, for exam-ple, that the pulsed energy driving the main drive motors be distributed in time—in other words, phase rotated. At room temperature, this can extend the life expectancy of the battery from less than 1 minute to just under 6 minutes under worst-case demands (all motors stalled, most systems working).

Since the state of charge of the battery can be indicated directly through measurement of its voltage, this becomes an extremely impor-tant parameter to keep track of. There are a number of ways of accom-plishing this, and they're all pretty simple, pretty inexpensive, and pretty effective.

The National Semiconductor LM3914, for example, was de-signed as a driver for LED bargraph and moving-dot displays. It in-cludes a voltage reference, a buffer amplifier, and a string of ten voltage comparators in an 18-pin DIP IC package. It can be configured as an expanded scale voltmeter in the moving-dot mode, turning one output on at each of ten (or any subset of ten) input voltage levels. See the National Semiconductor specification sheets and application notes for details and an actual circuit. Similar circuits have been presented in a number of magazines, including *Popular Mechanics*, *Popular Electronics*, and *Radio-Electronics*, as well as in *Engineer's Notebook—Integrated Circuit*

GLOBE-UNION ⊕
BATTERY DIVISION

5757 NORTH GREEN BAY AVENUE • MILWAUKEE, WISCONSIN 53201
414-228-2393 TWX #910-262-3084 TELEX 026-650

RECHARGEABLE 12 VOLT
UTILITY BATTERY
GLOBE PART NO. U-128

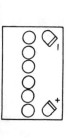

U-128

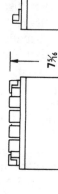

7⅝

7¾

5³⁄₁₆

DIMENSIONS (INCHES)

U-128 SPECIFICATIONS

1. Nominal voltage   12 volts (6 cells)

TYPICAL PERFORMANCE DATA

1. Reserve capacity (80°F)   28 Min.
   25 amps to 10.5 volts

1.4 amps (20 hr. rate) to	10.5	28 A.H.
2.3 amps (10 hr. rate)	10.3	23 A.H.
4.2 amps (5 hr. rate)	10.15	21 A.H.
15 amps (1 hr. rate)	9.8	15 A.H.
25 amps (30 min. rate)	9.6	12.5 A.H.

3. Weight — 21 lbs.

4. Internal resistance of charged battery — Approx. 10 milliohms

5. Maximum discharge current — 500 amperes

6. Operating temperature range
   Discharge — −76°F to + 140°F
   Charge — −4°F to + 122°F

7. Case material — Polypropylene

8. Vents — Pressure relief vents permanently attached

9. Grid material — Lead strontium

10. Sealed construction – batteries utilize a fully gelled electrolyte – will not leak or spill even if left upside down for extended time periods.

11. Terminal – "L" blade with clearance hole to accept ¼" bolt at negative and .350 inch sq. cutout at positive.

2. High rate discharge @ 80°F
   100 amps to 7.2 volts — 5.7 Min.

3. Low temperature discharge 190 amps @ 0°F
   5 sec. voltage — 7.7
   30 sec. voltage — 7.4
   time to 7.2 volts — .75 Min.

Discharge Curve

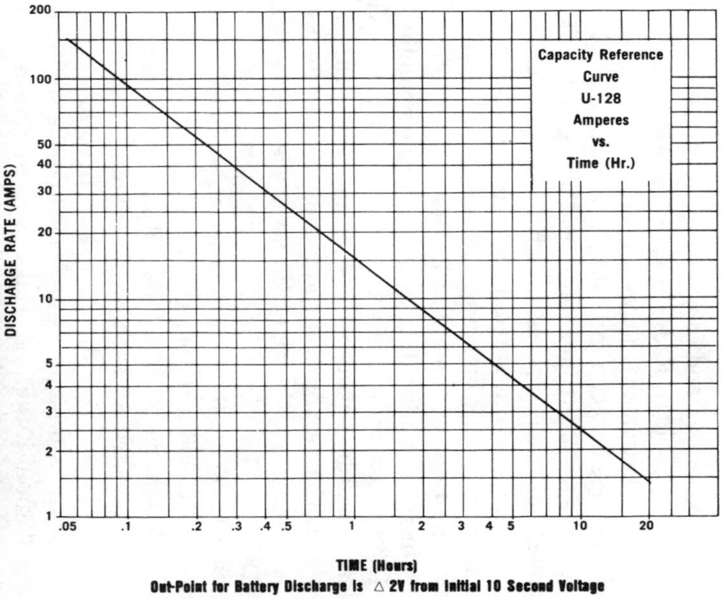
Capacity Reference Curve U-128 Amperes vs. Time (Hr.)
DISCHARGE RATE (AMPS)
200
100
50
40
30
20
10
5
4
3
2
1
.05
.1
.2
.3
.4
.5
1
2
3
4
5
10
20
TIME (Hours)
Out-Point for Battery Discharge is △ 2V from Initial 10 Second Voltage

Fig. 10-2 *Data sheet for Globe-Union Model U-128 Gel/Cell®.*
(Courtesy of Globe-Union)

Applications, written and compiled by my crazy friend Forrest Mims and published and sold by Radio Shack (where the LM3914 also is available).

One extremely interesting circuit is the Intersil ICL8211 Micropower Voltage Detector/Indicator. This 8-pin DIP IC includes a 1.2 VDC reference, a voltage comparator, and an output buffer. Output current is limited to 7 milliamperes, which permits direct drive of a LED (as in an optoisolator) without requiring a current-limiting series resistor. The Intersil applications literature includes a schematic for a low voltage battery indicator with nonvolatile (no loss with power removed) memory, which takes advantage of a hysteresis capability designed into the IC. The *turn-on* voltage is set at some fixed point (which for a 12-volt battery might be between 7.2 VDC and 11.4 VDC, depending on how deep a discharge you want to permit before the android "knows" it needs recharging). The higher *reset* voltage is set to somewhere between 12 VDC and 14.4 VDC (the charging voltage for a 12 VDC gelled-electrolyte battery), so the battery must certainly be undergoing a recharge before the IC will turn its alert off.

A number of these level detectors may be used, with a different trip point selected for each, to allow your android some sense of judgment as to how severe the immediate need for recharging is.

Also, you might want to incorporate some number of level detectors without hysteresis: Those with hysteresis provide a memory of how low the battery voltage has dipped under load, while those without can indicate whether the battery voltage has recovered to back above this level.

One very simple approach to this problem uses optoisolators, limiting resistors, and Zener diodes. The sum of the Zener voltage and the LED turn-on voltage (normally between 0.7 and 2.5 VDC, depending on the LED) determines the voltage at which the optoisolator will turn on. This provides discrete system-compatible outputs at any selected number of voltages.

An alternative to the Zener diodes might be the Texas Instruments TL430, a 3-lead device that can be programmed with two resistors to perform like whatever value of Zener diode you need.

Another Texas Instruments device worth looking at is the TL489 5-Step Analog Level Detector. Like the National LM3914, this was designed to be used as a 5-point bargraph driver (although, to be fair, the TL489 has been out longer). This 8-pin DIP houses a voltage regulator and five comparators, switching the outputs on successively as the input pin voltage rises from 0 to 1 volt, each at a 200-millivolt interval. Used in conjunction with a TL430 and four resistors, you may be able to build a very compact multipoint charge sensor.

Let's set some guidelines for determining at what voltage the battery might be considered "dead."

Normally, at the end of a manufacturer-rated cycle, the battery will deplete to 1.75 VDC per cell, or 10.5 VDC for a 12-volt battery (nominal for 6-cell battery) after performing steadily at the rated current drain for 20 hours at room temperature.

Abnormally, under near short circuit conditions, the battery might deplete to 6.0 volts—not enough to keep a regulator supplying operating voltage for the memory alive.

The data sheet for the U-128 (see Fig. 10-2) suggests that the nominal 30-minute rate discharge (25 amperes) leaves 9.6 VDC in the battery. At 80°F, after producing 100 amperes for 5.7 minutes, it's 7.2 VDC. At 0°F, where the output voltage is lower than usual room temperature voltage anyway, while producing 190 amperes, the battery voltage reduces to 7.7 VDC in 5 seconds, to 7.4 VDC in 30 seconds, and to 7.2 VDC in 45 seconds.

Which means, by way of bold inference, that if we alert the android to an "emergency" need for recharging at 7.7 VDC and limit its current drain to 190 amperes (the four main drive motors near stall and the other systems not doing anything extraordinary, like turning on lights or waving arms), the android has about 40 seconds before it will reach "pass-out" voltage. This is even enough time to climb a stairway, which is the probable worst-case requirement.

Looking at the best-case condition, where the android is not required to move around every second of his "work day," doesn't spend a lot of time climbing stairs, if any, and has plenty of opportunity to "relax" and recharge, we might look at 10.5 VDC as a "might-as-well" hunger signal. Once the battery has reached this level, the android would charge up—casually—as the opportunity (in other words, lack of other tasks) presents itself.

At 10.0 VDC, the need to recharge becomes a little more serious. At minimum, the mapping subsystem should be consulted to determine whether or not a usable power outlet has been located within the room the android is then in. If not, finding one must become a high-priority task. If so, the task-in-progress should be evaluated to determine whether or not it can be abandoned in favor of plugging in.

At 9.5 VDC, the android should recharge itself immediately unless doing so would place life or property in danger, either because of the act of recharging or because of the nature of the task the android is then accomplishing. This is the lowest voltage at which the android should attempt climbing a stairway in search of an outlet—and may be *iffy* even so.

At 9.0 VDC, if the android has not located a suitable outlet, it should have the option of unplugging nonessential cords and plugging into their sockets. If this option is not accessible, it should enter an energy conservation mode, minimizing use of nonessential motors and subsystems, staying in one place, if possible, and periodically calling for help.

This sequence, as the others, should be overridden if damage to life or property can result.

At 8.5 VDC, you may wish to let the android exercise some extraordinary options—like memory-protect and complete power-down except for periodic calls for help, say every 1/2 to 4 hours.

At 8.0 VDC, even where damage to property may result, even where specific instructions might be ignored, the survival of the android is critically balanced. There is absolutely *no* chance that the android can climb a stairway once the battery charge has depleted more than a few tenths of a volt beyond this level.

At 7.0 VDC, the android "passes out," meaning it is unable to do much of anything beyond attempting to save its memory.

The tasks you intend to set for your android play a role in determining what its reaction should be at any given voltage level.

By the same token, each task you include in programming should provide for specific actions (or inactions) at each voltage checkpoint. Learned or heuristically adapted tasks also should be subjected to an internal regimen, which means in programming terms that action/inaction variables should be automatically incorporated and reexamined by an interrupt-executable subroutine each time a checkpoint alert occurs—probably at the overseer-processor level.

So far, we have examined our options in terms of battery selection and charge-point monitoring. We still have to look at connection to the rest of the circuit, connection to the outlet, how to find the outlet, how to qualify the outlet, and how to design the charger itself. Let's start inboard and work our way out.

In the case of the U-28 Gel/Cell, the negative terminal is an L blade with a clearance hole drilled to accept a 1/4-inch bolt; the positive terminal is an L blade with a 0.350-inch-square cutout. The easiest way of mating to these terminals is probably with battery cable lugs on #2 or #4 welding cable, a lockwasher, nut, and bolt. (Don't confuse the lug end, which is flat with a hole in it through which a bolt can be mounted, with the connector that tightens onto the battery post.) Alternatively, automotive switch-to-starter cables can be used. If your local automotive parts supplier can't find you what you need, contact Belden Corporation (2000 South Batavia Avenue, Geneva, IL 60134) or AMP Inc. (Harrisburg, PA 17105).

The choice of #2 or #4 welding cable for power connections from the battery is based on its flexibility and the peak current requirement of the main drive motors at stall—by far the largest portion of load requirement on the battery.

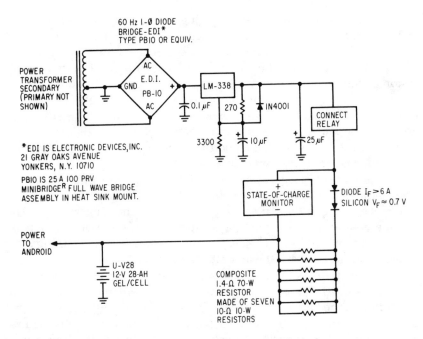

Fig. 10-3 *Schematic of voltage regulated, current limited power supply for battery charger.*

A smaller cable, while affording some savings in weight, can overheat under peak current demand conditions—dangerously hot—and can "burn up" some of the current the motors need.

Recharging the battery is not a technically difficult task but requires some knowledge of the way the battery accepts charge. Chemically, water and lead sulfate within the cells combine to reform the H_2SO_4 component of the electrolyte and the lead-oxide and lead of the plates.

Electrically, initially high currents surge into the battery, gradually increasing its voltage, in turn reducing the current inrush. An unlimited current source could recharge the battery very quickly but at the expense of drying out the water in the electrolyte and reducing the life of the battery. Normally, the battery could be good for 300 to 500 charging cycles—several years of light use, about a year of heavy use. If the battery is recharged often after only minor discharge cycles, over a thousand cycles are possible.

Globe strongly recommends constant voltage, limited current charging (see Fig. 10-3). A voltage of 2.4 VDC per cell (14.4 VDC for the U-128) permits optimum recharge time. Higher voltages can accelerate gassing, drying, and aging; lower voltages mean longer charge time and possible less-than-full recharge.

Given constant voltage charging at 2.4 VDC per cell, the best overall performance and life can be achieved at currents roughly 3 to 4 times the 20-hour rated current for the battery—4.2 to 5.6 amperes for the U-128. At this rate, the battery will have 80 percent of its capacity restored in less than 8 hours, 100 percent in less than 16 hours.

An unlimited current charger could restore full charge to a battery in 2 to 3 hours, but at some expense in battery life, charger weight, and expense and component requirements. Normally, this is not recommended, but there may be good reasons you need to give your android this fast-recovery capability.

Let's take a look at some extremes. Assume your battery has been depleted to a remaining voltage of 6.4 VDC, meaning an 8-volt difference between the battery voltage and the charging voltage. And assume you design your charging supply for current limiting at a maximum 6-ampere output. That means more than 50 watts through the transformer and nearly 50 watts into the battery.

Something between these two voltages (normally either a resistor or a lamp) helps ease the transition between the charger output circuitry and the battery for those millions of friendly little electrons the power company sends to do our dirty work. By calculation, this requires a 1⅓-ohm 50-watt resistor, which can be made up of a number of smaller-wattage resistors, if desired.

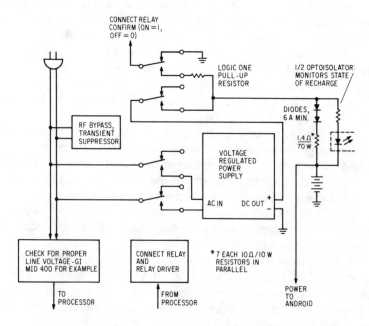

Fig. 10-4 *Schematic overview of battery and charger subsystem.*

Generally, well-filtered regulated dc supplies are preferred as chargers for these batteries. This helps us take advantage of the voltage drop across this resistor—an indicator of the current through the resistor, of course—as an indicator of when the battery has reached full charge.

Globe-Union literature suggests an end-of-charge current of 180 to 460 milliamperes for the U-128, yielding between 0.24 and 0.61 volts across the resistor. The voltage across the resistor can be monitored by any number of voltage comparators, including, for example, the National Semiconductor LM393, LM2903 series.

A number of adjustable voltage regulator ICs are now available that are capable of providing the output current the charger requires. These include the National LM 338; the MIVR-42050 from Micropac Industries, Inc. (905 East Walnut Street, Garland, TX 75040), a hybrid IC rated at up to 10 amperes and up to 34 VDC; and many others.

The design of the dc supply used as the charger is straightforward and can be either built or bought, depending on your preferences. The requirements are for 14.4 VDC, well regulated, a 5.6- to 6.0-ampere capability and, most simply, a 1.3-ohm (approximately) 50-watt resistor between the supply and the battery to both limit and monitor current. Your decision should be influenced by cost, availability, size, and weight.

But we're not done yet.

Although we haven't talked about it, we should consider the sequence of events in plugging in for a charge. We need to decide what happens to the android systems, what the battery connects to, what it disconnects from, and so on (see Fig. 10-4).

First, it should be obvious that we can't afford to remove power from memory or the android would literally forget what it's doing, as well as where it is and what it should do next.

Second, since keeping the battery connected even during charging means isolating hum and power line noise from the bias and data lines, this strongly suggests filtering of the ac line input, transient suppression, and bypassing; plus independent filtering and bypassing at the lower-voltage regulators for the various circuits within the android.

Third, it will do us some good to confirm an adequate primary voltage before activating the charger, since too low a voltage at the primary can mean an insufficiently high voltage at the battery and actually pull down the battery instead of recharging it.

Two remedies are suggested here. The first is a unique device from General Instruments Optoelectronics, the MID400, a power-to-logic interface. A resistor controls the turn-on voltage of a LED, which is optocoupled to a logic-level driver (see Fig. 10-5). The result is a logic output when the line voltage is adequately high.

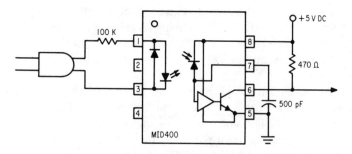

Fig. 10-5 *The MID400 ac line monitor optically isolated interface device, two resistors, and a capacitor provide a logic level output (logic low) whenever the ac line voltage exceeds some resistor-determined minimum (here about 80 VAC RMS). The circuit is purposely designed to react "sluggishly" so half-cycle zero crossings are not counted as outages.*

The second remedy involves a minor redesign of our charger power supply. A diode in series with the lead from the charger and resistor to the battery can assure that the battery won't discharge into the charger but requires that the forward voltage of the diode be added to the charger output voltage. Depending on the diode, this might be from 0.4 to 0.8 VDC, requiring 14.8 to 15.2 VDC from the supply.

Design against the lowest rated forward voltage drop of the diode for the range of currents encountered. For example, the 80SQ series 8-ampere Schottky Power Rectifiers from International Rectifier (Semiconductor Division, U.S.A., 233 Kansas Street, El Segundo, CA 90245) demonstrate forward voltage drops of from 0.3 to 0.6 volts. Using the lower figure, the charger power supply output would have to be 14.7 VDC. If the diode drop (the value that increases with current and decreases with temperature) were to rise to the higher figure, the effective output of the charger would be reduced from the nominal 14.4 VDC to 14.1 VDC.

Now let's put a few things together. Battery charging current demand is highest at the beginning of the charge, when battery voltage is lowest. Diode forward drop is highest at this point, too, which means that even though the charging voltage on the battery is marginally lower at first, the voltage and current are both high enough to permit little or no loss in charging efficacy.

Once the battery has accepted additional charge, the charging current reduces, lowering the diode current and its forward voltage drop, and bringing the charging voltage back up to its nominal 14.4 VDC level.

The additional piece of a volt produced by the diode's forward drop lets us try a simple approach to determine when battery charging has been completed. A current limiting resistor in series with a LED can be connected with the anode connected to the power supply output, the cathode connected to the battery. This LED will be on as long as the battery is charging but not at or near full charge—at this point the voltage drop across it is too small to keep it lighted. If the LED is packaged in an optoisolator, the result is a logic-level output.

Before we take a look at the whole sequence of events entailed by the battery-charging task, let's add one more piece to the puzzle.

The android needs some means of finding outlets to plug into. Live outlets. Visual recognition is a difficult task at best and one some of us may never quite accomplish.

The alternative is an experimental approach I call a "Mainsfinder." This is a simple directional antenna, FET (field-effect transistor) input amplifier, and level detector. My initial experiments are based on an earlier signal tracer designed with a JFET and a packaged audio amplifier, a Zener detector, capacitor integrator, and npn transistor output driver. The antenna was a piece of double-sided printed circuit board stock.

The mainsfinder picks up hum fields, and when the detected amplitude is high enough (as determined by the Zener value), it switches on the output. The antenna acts like most dipoles, with sharp null characteristics.

A better approach can certainly be developed with a little more time and attention on my part—or yours. One area I have yet to try is a directional antenna designed with an aluminum cup and a size-matched double-sided PCB disk inside. The cup and inside surface of the PCB are grounded; the outside disk surface is connected directly to the FET gate.

Also, BiFET op amps might help reduce parts count and complexity. As I've indicated, this approach is still very experimental, but nevertheless very promising.

About the only components we've left out of the story at this point are the power cord reel, the plug handle, and the connect relay.

The cord reel is a fixed drum with a rotating guide arm, which is motorized, and provides a set of cord guide rollers which orbit the drum (see Fig. 10-6). This permits fixed-position stowage for the cord, easy access, and no dangling or tangling.

The plug itself—that little take-it-for-granted thing we grab and plug into the socket—could present a difficult grasp-and-manipulate problem for the android. A simple T-shaped handle makes the job easier. It should be made of an insulating material, such as plastic or rubber. A flat, rectangular cross arm 3–4 inches wide, ¾–1 inch thick, and 1–1½ inches deep provides a firm gripping handle that won't rotate; the upright section of the T should be narrow enough to easily fit be-

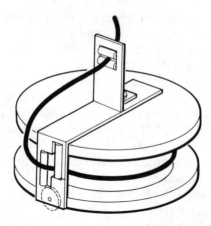

Fig. 10-6 *Mechanism used to wrap power cord around fixed flanged drum. Drive motor inside drum drives guide arm through wrapping orbit of two orbiting roller guides. One is friction driven by flange to help propel cord. Optionally, a small wheel (shown with dashed lines) could help reduce friction.*

tween the android's fingers. The goal is to provide something the android can grasp easily and firmly with a minimum of hassles and guide easily into the outlet.

Finally, the connect relay is a 6-volt 4-pole double throw relay capable of switching 6 amperes or more (which in the "real" world probably means a 10-ampere contact rating). This relay switches the 110-VAC power cord to the power supply and the power supply to the battery; two poles are used on the mains, one on the output, and the last one provides a confirmation to the logic circuits.

Now, at long last, let's look at the sequence of events that occurs when the android decides it's time to recharge.

1. The overseer processor is alerted that the state of charge on the battery requires attention. It examines the other tasks and requirements the android is facing and decides to service the need to recharge.
2. The mapping processor is consulted for the coordinates of the nearest power outlet. If any are found, they are examined for availability of sockets and once one is found, the plug-in sequence begins. If none is found, the android begins searching for one, using the mainsfinder and all the other senses available to it until one is found. If none is found, the maps for adjacent rooms are consulted. If none is found in any adjacent room, the android calls for help or, depending on the state of charge, enters and searches the other rooms, beginning with those on the same floor and descending stairs in favor of climbing them.

3. The android positions itself as close as possible to the outlet, preferably facing it squarely. The mapping, ranging, visual, and calculating processors cooperatively provide relative coordinates of the outlet and translate these into commands for the arm and finger motors. This is a continuing process during the following maneuvers.
4. The android unreels the full length of power cord while holding the plug handle and pulling up the slack with its arms.
5. The android holds the plug in its hand in a predetermined position in front of the visual system camera(s) to confirm the proper grasp attitude; if incorrect, the second arm and hand are used to correct it.
6. The android positions the plug directly in front of the outlet with the prongs directly opposite their mating slots. The visual system confirms proper orientation or directs the arm and hand motors to achieve the proper orientation.
7. The android plugs in.
8. Proper line voltage level is checked. If not present, the android unplugs and plugs in again. If still not present, the android moves to the next available outlet and tries the entire sequence again. If present, the connect relay is engaged and charging begins.
9. The state of charge is monitored during the charging process. Once the battery has received full charge (or if the overseer processor commands that the charging cycle be interrupted), the unplug/disconnect cycle begins.
10. The connect relay is released.
11. The android pulls the plug from the wall.
12. The android slacks the cord as the cord reel guides it back around the storage drum.

That completes the charging cycle. And this completes our look at the battery and charging system.

11
Motor Juice Concentrate

If it weren't for a couple of very specific technological break-throughs—VMOS power FETs and pulse motor drive—it would be all but impossible to design an android capable of stair climbing with any reasonable amount of service life between recharge cycles. That's because the forward voltage drop in a conventional transistor becomes a significant power consumer at high drive currents.

For example, four motors drawing 50 amperes at stall through conventional H-configuration drivers both reduce the (nominal) 12 volts available to about 10½ volts and force the eight transistors to convert nearly 300 watts into heat. (Fanatical supporters of bipolar transistor drivers wishing to conserve energy might convert this heat into steam and drive a small on-board generator.)

Even relay-reverser drives using a single transistor at each motor cut between 1/2 and 1 volt from the available supply and eat up over 100 wasted watts.

Compare these figures, for example, to the International Rectifier IRF 150 VMOS FET. At 28 amperes, these devices exhibit an on-resistance of 55 milliohms. Two of these devices can be paralleled to permit the required 50-ampere drive, drawing 25 amperes each.

But wait a minute! That means a voltage drop of 1.375 volts and current consumption of almost 70 watts. Is 30 watts such a big deal?

Probably not by itself. But VMOS FETs can be driven by nanoamperes, even directly by CMOS, whereas bipolar transistors capable of delivering these output currents require amperes of drive through additional drivers, which themselves waste power.

All of this so far is intended to intrigue you so much that you'll put up with a lengthier introduction to VMOS pulse motor drivers. If you'll permit, I can help you fall in love with them, too.

VMOS FETs are a special class of MOSFET transistors designed specifically for high current drive capability. Like other FETs (and, for us old timers, like vacuum tubes), these are high input impedance devices, requiring only very small currents and responding much more directly to variations in voltage.

VMOS FETs are majority carrier semiconductor devices, bipolar transistors are minority carrier devices. The base, collector, and emitter of a bipolar transistor correspond to the gate, drain, and source of a VMOS FET (and correspond most directly for an npn transistor and an n-channel FET; pnp, and p-channel FET). In the transistor, base-emitter current flow produces collector-emitter current flow, amplified by the gain of the transistor. In the VMOS FET, gate-source voltage produces drain-source current flow.

Because there is no direct connection between the gate and the source in a VMOS FET, theoretically there should be no current flow into the gate; in practice, however, there is a small leakage current—in the order of nanoamperes! As a result, the dc current gain is on the order of (ready?) a billion.

Like vacuum tubes, dc current gain isn't a very meaningful parameter to use with VMOS FETs, but *transconductance* is. Transconductance is a measure of the change in drain current for each 1-volt change in gate voltage.

A family of drain-current versus drain-source voltage curves for various given gate voltages can tell a lot about what VMOS FETs are all about (see Fig. 11-1).

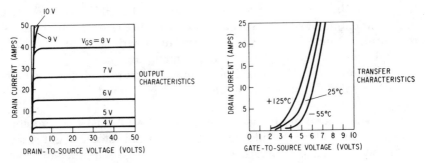

Fig. 11-1 *Key characteristic curves for typical high current n-channel VMOS FET.*

For any one given gate voltage, its curve exhibits two regions of interest: one (at lower drain-source voltages) is a "constant resistance" region in which the drain current increases nearly linearly with increasing drain-source voltage, the other (for drain-source voltages beyond a pinch-off voltage) is a "constant current" region which the output (drain) current remains nearly constant even as the drain-source voltage is increased.

The family of curves is such that as the gate voltage is increased, the resistance of the constant resistance region becomes smaller and the level of constant drain current increases.

The limiting parameter for the VMOS FETs output current is its on-resistance: the lower its on-resistance, the more current it can conduct and the lower the voltage dropped while doing so.

The family of curves we've been discussing illustrates another significant characteristic of the VMOS FET: up to a threshold voltage of a few volts, increasing the drain-source voltage has little effect (nearly none) on the drain current.

There's another advantage to these little darlings—switching delay times are extremely small, on the order of a few nanoseconds. This is because, as noted, the VMOS FET is a majority carrier device, meaning that its charge carriers are not controlled by the physical injection and extraction or recombination of minority carriers, as in bipolar transistors; rather, they are controlled by electrical fields. Unless you plan on baking your own, you'll be most interested in the comparative result of this advantage, switching times from about ten to several hundred times faster than bipolar transistors.

Ever had a transistor die because of thermal runaway? In a bipolar transistor, its output current increases with temperature, and unless externally limited or controlled, keeps right on doing it until it cooks or conducts itself to death.

In VMOS FETs, current draw *decreases* with temperature. So if the current drain at any particular point within the FET increases, that particular chunk of silicon heats up and the higher temperature reduces the current. Voila! No hot spots or current crowding because current throughout the chip equalizes itself automatically.

Bonus time! This same phenomenon occurs with devices connected in parallel. Current is automatically shared and ballasting resistors aren't required.

There is no mechanism inherent in a VMOS FET that prohibits it from withstanding its full rated voltage and full rated current simultaneously. Inductive loads, for example, don't present a problem. Compare that to your experiences with bipolar transistors and you'll come another step closer to the VMOS fan club.

A little practical advice from the manufacturers suggests a few very simple precautions when paralleling VMOS FETs. Because these devices are so fast, ferrite beads or small resistors—something between 100 and 1,000 ohms—in series with each gate will suppress spurious oscillations (which could be several hundred megahertz).

For switching applications, like our pulsed motor driver, it's important to switch the VMOS FET on as hard as possible. To best accomplish this, a 10K pull-up resistor to the +12 VDC supply at the gate input lets a CMOS output provide all the drive necessary—or open collector TTL, if you prefer.

One manufacturer, International Rectifier, has stated that all of its VMOS FETs will conduct their full rated continuous drain current with less than 10 volts from gate to source and that the threshold voltage will always be greater than 1 volt, assuring a cutoff condition with no special reverse biasing required. Their IRF-150, cited earlier, for example, has a rated threshold between 1.5 and 3.5 volts and requires only about 8 volts at the gate for full output.

Before we move on to some of the specifics of design (which aren't really complicated now that VMOS FETs are here, thank goodness), here are some good sources of information about these particular devices.

Siliconix, Inc. (2201 Laurelwood Road, Santa Clara, CA 95054) offers a *VMOS Power FET Design Catalog*. International Rectifier (233 Kansas Street, El Segundo, CA 90245) offers *Power MOSFET Application Notes*. These manufacturers and others offer specific device data sheets. Write the above plus Supertex, Inc. (1225 Bordeaux Drive, Sunnyvale, CA 94086); Hitachi America, Ltd. (Electronic Devices Sales and Service Division, 707 West Algonquin Road, Arlington Heights, IL 60005); Intersil, Inc. (10710 North Tantau Avenue, Cupertino, CA 95014); and Texas Instruments, Inc. (P.O. Box 225012, Mail Station 34, Dallas, TX 75265).

In addition, the June 7, 1979 issue of *Electronic Design* included an excellent discussion of the International Rectifier HEXFET™ VMOS power FET, by far the most capable devices to date.

By the time this appears, more manufacturers should have announced p-channel VMOS FETs, which will permit an H-configured reversing motor drive using VMOS FETs only, without relays.

By the way, when reverse biased, VMOS FETs act like forward-biased diodes—an interesting phenomenon when you consider that motor designs require a flywheel diode to conduct the inductive kickback of the drive pulse from the motor. This tidbit comes courtesy of Ralph Waggitt, Product Marketing Manager for the Semiconductor Division of International Rectifier. This delightful Englishman also predicts that increased competition will be bringing lower prices—probably by the time you read about it here. As of this writing, a 28-ampere 60-volt VMOS FET has a 1–9 piece suggested resale price of $70.00. Compare the prices now.

We also should see some new packaging. The TO-3 package of most current devices means too much capacitance to take full advantage of the high-speed capabilities of VMOS; the TO-220 packages of most of the others (those that aren't in TO-3s) aren't built for high current. Here, you have the advantage. All I have is a cloudy crystal ball; you can check out actual data and specification sheets.

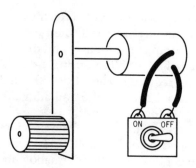

Fig. 11-2 *Crank the motor with switch across motor leads off; then continue cranking with switch on. This is equivalent to placing an infinite load on a dc generator. The result is inertia—braking the motor.*

We haven't said too much about braking or speed control, which are both vitally important topics. We'll investigate speed control through pulse width/pulse duty cycle varying controllers in a short while. Right now, we're going to look at a little trick that lets us accomplish braking without a brake and without consuming power. Interested?

This one is so good that I really want to walk you through a little exercise and let you feel for yourself how it works.

You'll need a small permanent magnet dc motor—something from your junk box, perhaps. Clamp it down somewhere and rig some sort of a crank handle to its shaft. Also, wire a switch across the two motor leads—that's right, you'll be turning on a short between the leads (see Fig. 11-2).

Now, with switch off, crank the sucker. Keep cranking, if you can, as you turn the switch on. Intriguing?

Okay, now take the crank off the shaft and replace it with a flywheel—some kind of massive, radially symmetric load. And get a double-pole double-throw switch. Hook a battery up to one side, a short up to the other side, and the motor in the middle—you'll be switching the motor from the battery to the short. Mark one spot on the rim of the flywheel with a piece of colored tape or a marker (see Fig. 11-3).

Start the motor and let it run. Note how long it takes the motor to stop. Then do the same thing, but cut the short first. See how the motor coasts to a stop. Unquestionably, the short loaded the motor.

Let's get fancy. Take a second double-pole double-throw switch, wire the motor to one side, cross leads to the other side, and wire the middle to the center of the first switch. Last but not least, wire the single-pole single-throw switch to the terminals that have the remnants of the short circuit you cut open.

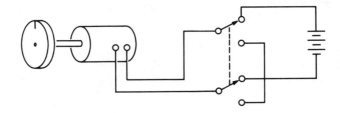

Fig. 11-3 *This arrangement switches motor between battery and short. Tape or mark on flywheel permits easier observation of motor speed. Motor starts and runs when connected to battery and brakes quickly to a stop when switched to the short.*

What you have is a forward/reverse switch, a run/stop switch, and a coast/brake switch that provide complete start/stop and direction control (see Fig. 11-4). The run/stop switch connects either the battery or an openable short to the motor windings. The coast/brake switch selects whether the motor windings will be shorted (causing braking to a stop) or open (permitting coasting to a stop) when not connected to the battery. The forward/reverse switch reverses the motor by reversing its leads. All this can be accomplished cheaply and easily.

It isn't going to be quite so cheap and easy to do the same thing for the android. We'll be using VMOS FETs to drive relays, under processor control, other VMOS FETs to provide the pulse drive to the motors for speed control, and lots of little feedback tricks to keep things where we want them—or at least to know how wrong they've gone.

For example, in an emergency stop situation, reversing the motor during narrow width pulses after using the short circuit trick to begin braking can be as effective as mechanical braking.

Normally, you see, the motor could be either ramped to a slower speed during a planned trip—meaning the motor is driven to successively slower speeds by using narrower and narrower pulse drive, reducing the android's momentum—or it could coast to a stop, or the short circuit could be switched in. But to speed up the braking, when the control relay switches the motor to the short, the direction relay can

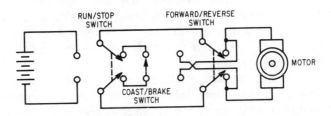

Fig. 11-4 *Just three switches provide complete start/stop and direction control.*

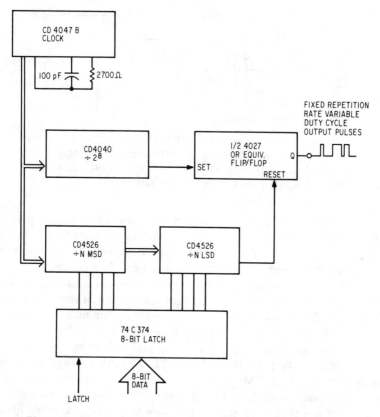

Fig. 11-5 *Circuit shows way of allowing data to vary pulse width from 0 to 100 percent over 256 steps at fixed pulse repetition rate.*

reverse the motor leads, and immediately the pulse drive circuits can begin pulsing the motor against the motion of its momentum.

But whoa! We've brought up a lot of subjects we haven't really begun to cover, like pulse motor drive, the motor relays, the flyback diodes, and feedback. Let's take these things one step at a time. (Notice how cleverly the author pretends to have gotten ahead of himself so he can cover all these things one at a time, including a sense of overall perspective, without having to make them flow into each other until everything's just about finished! I only bring it up because that's exactly what the task of designing an android can be like—lots of little pieces you know you need, but you're not really sure which should come first, so you start where you can.)

Traditionally, motor speed has been controlled by varying the motor voltage. For dc motors, this was most often accomplished with a rheostat or some other form of variable resistance tapping down from a higher voltage. Later, the idea of a variable dc supply gained favor.

But control of motor speed and acceleration with variable voltage control is difficult. This is especially true when starting from a stop. Normally, voltage is increased until sufficient power is available to break loose of the inertia keeping the rotor in position, but that same breakaway motor voltage, sustained, quickly accelerates the motor to a very high speed—something like popping the clutch in a stick-shift car.

Once started, going slow can be a problem, too. Using variable voltage, you have to get the motor started, drop the voltage as soon as it has, and make sure it doesn't drop too far or else start over and try again. Problems, right?

The solution is in the nature of motors themselves. A motor exhibits one particular parameter—its *time constant*, which is the time required to reach 63 percent of its rated speed (in the case of its *mechanical* time constant) or rotor current (in the case of its *inductive* time constant)—which permits some of the effects of transients appearing at the motor terminals to be averaged over time.

Pulse-width modulation (also known as PWM) feeds voltage and current to the motor in chunks, so to speak. Generally, constant-amplitude (voltage) pulses are used. Of the several variations, we'll be discussing the one in which pulse frequency (repetition rate) is fixed and the pulse width is varied.

Each pulse, then, provides a given-size "chunk" of power, usually at the full-speed rated voltage for the motor and with whatever current the motor draws under load. Each pulse provides a little push, a little bit of "oomph" or boost to kick the motor around. A series of little pushes drives the motor slowly; a series of big pushes drives it faster.

(Believe me, this may sound simplistic, but it beats heck out of going through rigorous theories about motor dynamics and achieves the same end result.)

With PWM, a series of narrow pulses are enough to start and maintain slow speed. Increasing pulse width increases speed (and available power).

Frequencies as low as 5 to 200 hertz are enough to assure smooth, continuous motion; higher frequencies are fine, too. An excellent discussion of the subject, including a schematic for a 555-based PWM controller is included in Rodney A. Kreuter's article, "Controlling DC Power with Pulse-Width Modulation," from the June 1979 *Popular Electronics*.

A digital pulse-width modulator can be built or breadboarded very simply. A clock drives both a presettable-countdown chain and a fixed divider; the chain resets a flip-flop, and the fixed divider sets it.

The schematic in Fig. 11-5 shows a CMOS version of this circuit, using a CD4047B low power astable multivibrator (you could easily substitute the Intersil 7555, a CMOS version of the 555, an oscillator built from gates, or a system clock), a CD4040B 14-stage ripple carry binary divider, two CD4526B programmable divide-by-N 4-bit binary counters, half a CD4027B dual J-K master/slave flip-flop with set and reset (or use a quarter of a CD4043B quad Tri-State NOR R/S latch and exchange set and reset lines) and an MM74C374 octal D-type flip-flop.

The '374 latches in data for the cascaded programmable dividers. The eighth stage of the 4040 sets the flip-flops ($1/2$-4027 or $1/4$-4043) every 256th clock pulse; its output remains high until the Nth count from the 4526s resets it—N is a number between 0 and 255.

The output, of course, eventually reaches the VMOS FET power driver for the motor being controlled. We'll talk about the route in a moment.

First, though, let's stub our toes and gain a little brainpower. We know about (or can forecast) four main drive motors, a trunk lean/tilt motor, a trunk rotate motor, a head rotate motor, a bunch of arm motors, a battery charger cord reel motor, and we know there are going to be more. Are we going to repeat this $5^{1/2}$-IC circuit a few dozen times?

No need. A one-chip microprocessor, a PIA, and maybe a buffer should be able to handle as many as 16 motors at a time. Tell it what speed (data) to drive what motor (address) and let its programming loop through the requisite numbers of counts. Each PIA line can drive a motor's VMOS FET power stage.

That leaves the braking and directional control tasks—remember our toggle switches? We'll be using a state-of-the-art component to accomplish each part of this task—good old electromechanical relays.

Since we've been talking about using two motors on each side of the drive (one for each pulley or wheel set)—and have been considering motors with stall currents on the order of 50 amperes—does that mean using one relay on each side to make braking and reversing happen or two? The question is really whether we have to specify a 50-ampere relay or a 100-ampere relay.

That decision has pretty much been made for us. Either one would be hard to find and quite costly. So we'll just have to do the best we can with what we can get—then cheat (somehow) to make it work. Stay tuned, ye unfaithful, for the secret answer to this mystery—but first these words from the specification sheets.

Square D Company (P.O. Box 472, Milwaukee, WI 53201) type CO-15 and CO-16 relays are rated up to 30 amperes at up to 240 volts and are available in DPDT contact configuration.

Omron Electronics, Inc. (650 Woodfield, Schaumburg, IL 60195) offers two likely candidates. Model MG2-UA-DC12 is a DPDT relay rated 30 amperes at 120 VAC with 12-VDC 57-ohm coil. It pulls in at

9.6 VDC, drops out at 1.2 VDC. Maximum initial contact resistance is 25 milliohms. It requires a maximum of 40 milliseconds to pull in, 50 milliseconds to drop out. Temperature range over which it's designed to operate is −10 to +50°C. Its mechanical life rating is a minimum of five million operations; electrically, under the maximum rated load (30 amperes), it's rated for a minimum of a half million operations. With a hundred operations a day and everyday use, that's 14 years of service.

Mechanically, the Omron MZ2-UA-DC12 is laid out differently than the MG-series model, but most of its specifications are identical— with a few exceptions. This MZ-series relay is rated for only half the life of the other, both mechanically and electrically. It has a lower contact resistance, 45 ohms, requiring 3.6 watts versus 2.5 watts to hold the relay in. On the good side, it releases a hair faster—40 milliseconds maximum instead of 50.

Midtex, Inc. (Aemco Division, 1650 Tower Boulevard, North Mankato, MN 56001) offers a type 302-11A200 DPDT power relay with silver-cadmium oxide contacts rated 30 amperes at 277 VAC (resistive), a million cycles mechanical life, and hundred thousand cycles at full load. Contact resistance is initially 200 milliohms, finally reaching half an ohm (rated at 1 ampere and 20 volts—and no, afraid I don't know how to translate that into a final resistance at the end of its full-load electrical life). Its 12-VDC coil pulls in at 9 VDC, offering 71-ohm resistance for a 2-watt power requirement. Rated operating temperature range is from −55 to +80°C.

Magnecraft Electric Company (5575 North Lynch Avenue, Chicago, IL 60630) offers Model W199X-12, a DPDT 30 ampere (at up to 300-VAC or 28 VDC resistive load) with a 12-VDC 70-ohm 2.0-watt coil that pulls in at 9.6 VDC. The contacts are gold flashed silver alloy. Suggested U.S. resale price is under $10 at this writing.

There are other manufacturers and other relays available, of course. You are well advised to conduct your own research. But you're doubtless still wondering how a 30-ampere relay can be expected to make and break a 50-ampere current.

The answer is that these relays cannot be expected to make and break 50 amperes, but when required, they can be expected to conduct it. Let's examine this requirement in "real-life" perspective.

Consider, for example, that 50 amperes is the motor start and stall current. Once the motor has started and unless it meets a sizable load (yes, like gravity when climbing stairs), the required current is much, much less. So during the *majority* of its required operations, the relays will be conducting considerably less current—normally less than 5 amperes.

Consider that the relay will never have to conduct a 50-ampere current for long because the battery isn't capable of delivering a 50-

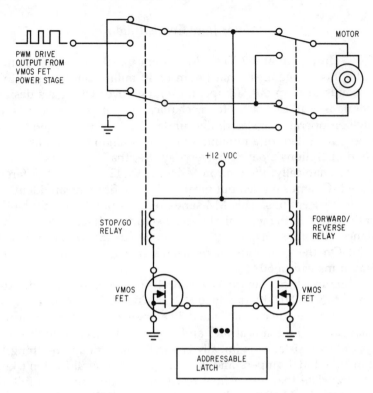

Fig. 11-6 *Configuration of power-handling motor control relay circuit.*

ampere current for long—not when you consider that the load will occur on at least two and probably four motors and their associated relays.

And consider that with the proper control, the relays can be permitted to pull in before any current is applied and that current can be cut before the relays release. That's how we can get around the requirement to make and break too high a current for the relay. It also removes the danger (to both contacts and combustibles) of arcing and sparking during make-or-break switching.

We've seen switching times of about 40 to 50 milliseconds. If we allow 80 to 120 milliseconds between switching the relay and beginning or ending pulse drive through the VMOS FETs, this should be ample time to allow for both contact switching and contact bounce.

A monostable multivibrator can be used to alert the controlling microprocessor that the wait period has transpired, or a real time (fast) clock can be read and compared, or cycles can be counted—all depending on what signals are handiest to your system and the microprocessor you choose for motor control. For example, if a 1,000-hertz clock is used to drive an 8-bit counter, the microprocessor can read the counter at the beginning of the wait period, go on and do something else, sampling the clock from time to time, and when the count is equal to the initial count minus 128 (give or take a few LSBs) know that enough time has passed. Or a 100-hertz clock can drive the interrupt line, causing an internal counter to be incremented by one, and when the count equals 8 (for 80 milliseconds, but you can juggle the figures to suit your own requirements), both accomplish the action and disable the interrupt line. There are lots of options, so take your pick.

The only link we haven't discussed is the relay coil driver. The procedure here that requires the least hardware involves addressable latches. Data is entered into a specific location in one of the these latches when that location is addressed and the chip is enabled—both of which a microprocessor can easily accomplish.

The CD4723 is a dual 4-bit addressable latch; the CD4724 and CD4099 are 8-bit addressable latches. There are many others available from many manufacturers, and you are again invited to consult either the data books of your favorite manufacturer or the contents of your own junk box.

These addressable latches provide a memory function independent of the microprocessor, and once the addressed data has been entered it maintains the selected state until altered.

The output of these latches can be used to drive a small, inexpensive VMOS FET (see Fig. 11-6). Since none of these relay coils requires more than 400 milliamperes, even under reduced voltage, a 1/2-ampere or 1-ampere VMOS FET is more than suitable. Today's inexpensive FETs are rated from 1 to 2 amperes—more than enough for this requirement—and you may be able to run them with no heat sink, depending on the requirements of the specific device.

The last subject remaining in this discussion now is the diodes needed to protect our circuitry against inductive kickback—both from the motor and from the relay coils. These are becoming easy to find.

Fuji Semiconductors (imported and distributed by Collmer Semiconductor, Inc., 4100 McEwen, Suite 244, Dallas, TX 75234) offers type ERG81, a 50-ampere 40-volt Schottky Barrier Rectifier. Varo Semiconductor, Inc. (P.O. Box 40676, 1000 North Shiloh, Garland, TX 75040) has a 60-ampere 45-volt Schottky, type VSK51. International Rectifier (Semiconductor Division, 233 Kansas Street, El Segundo, CA 90245) offers 50HQ045, a 50-ampere 45-volt power Schottky rectifier.

The advantage of Schottky rectifiers is their extremely fast recovery time—a must for the fast flyback spikes that inductively kick back from the motor windings and relay coils.

Have you digested everything? Once you have, you're ready to put together your motor base and drive it—controlling it with a switchbox on a tether (if you want and your microprocessor isn't ready yet). After all, android builders are born tinkerers. And now, at least, you know enough about the things you'll be tinkering with to go ahead without destroying them—or yourself.

12
The "Other" Motors

From early on, we've made it an absolute requirement that in order to be considered an android, our critters have to be capable of manipulating things in their environment. This requires motors—lots of them—used in clever ways to provide arms, hands, hips, and other anatomical equivalent motor joints. All of this is much easier said than done. This chapter will provide some advice on the doing.

Making the motors move, while a problem, is not the biggest problem we face. We must precisely control that movement. We must know the result of any movement and the current position of the motor shaft. Position is a much more significant factor than speed, in this case—none of the motors we'll be talking about will produce an end rotation greater than 540°.

For the time being, we can get the farthest the fastest by speaking generally in terms of approaches to solving these problems. I promise, though, to get down to cases soon.

Broadly speaking, our topic is defined as *servo* systems. A servo (again broadly speaking) is any system that adjusts itself to comply with an external command.

In a common simple servo circuit, a motor and a potentiometer are mechanically linked. The potentiometer acts as a voltage divider, providing a unique voltage for each possible position. The external command is provided by a second potentiometer, also acting as a voltage divider. The two voltages provide the input to a comparator; the comparator's output switches the motor on (and may or may not select forward/reverse motor direction, depending on the requirements of the specific application) and keeps it on until the motor shaft drives the potentiometer into a position where its output voltage is equal to the output voltage at the command potentiometer. At that point, the motor is switched off by the comparator.

Several phenomena occur that complicate this task. First, motors tend to overshoot a target value that is too tightly specified. For this reason, servo designers usually incorporate a *deadband*. The deadband is a small error range in which the comparator is forgiving; only an error that is greater than the deadband range will cause the servo to respond. The term *error*, as used here, corresponds to the difference between the commanded position (or voltage or signal) and the actual one.

Second, in a system in which motor direction is reversed as appropriate to error direction (which, since it is usually represented by a voltage or current, is sometimes referred to as *error polarity*), one manifestation of a too-small deadband or a too-fast motor is *hunting*. Hunting is a particular kind of motor behavior as it overshoots in one direction, reverses and overshoots in the other, reverses again and continues to swing back and forth around its target. You will probably encounter hunting in some of your early trials. Depending on how big the overshoots are, how small the deadband, how quick the motor, and how large the load, hunting may or may not decay. Occasionally, hunting is designed in purposely (usually by incorporating a negative-value deadband) to accomplish a back-and-forth scanning action.

Third, some systems are designed with *damping*, which can be either electrically or mechanically implemented. One way to provide damping is to make motor speed proportional to error voltage, so the smaller the error, the slower the motor goes. This can be accomplished either continuously or within stated error ranges. Another option is to permit hunting but to slow the motor each time the servo reverses.

While the parameters of servo design are fundamentally analog in nature, there is no reason not to accomplish servo control digitally. Indeed, today's most advanced servo systems are entirely digital in execution.

For example, Vernitech (a division of Vernitron Corporation, 300 Marcus Boulevard, Deer Park, NY 11729) manufactures a number of linear and rotary position encoders. These include several different types. Incremental encoders provide pulse outputs for every so many degrees (or fractional degrees) of rotation. Absolute encoders provide a unique output code for any rotational position. An incremental encoder would be used, for example, to count motor rotation from a known reference or to provide tachometric (speed measuring) data. Absolute encoders are used where the position of the shaft must be precisely known instantly. Vernitech also offers potentiometers, which we'll look at shortly.

Disc Instruments, Inc. (102 East Baker Street, Costa Mesa, CA 92626) offers in their GC-30 series of absolute encoders a 10-bit resolution Gray Code format that resolves to ±3 minutes of arc. Gray Code is a manner of binary encoding often used in transducer applications; the Gray Code changes in no more than one bit position between any two adjacent codes, which is theoretically less traumatic for the circuitry that has to recognize changing code. By contrast, BCD and binary codes can change in as many as *every* bit position between adjacent code values. For slow rotational speeds and fast controlling circuitry, the advantages are moot.

Trans-World Instruments, Inc.®, Optical Encoder Division (700 East Mason Street, Santa Barbara, CA 93103) offers Number GS-3406 10-bit Gray Code modular absolute optical encoder. A subminiature incandescent lamp rated for a hundred thousand hour life shines on a photocell but is interrupted by the selected code disk. (In addition to Gray Code, binary and BCD disks are available.) A 741-type op amp and LM339 comparators provide TTL-compatible outputs with a nominal fanout of three standard TTL loads. This is typical of designs for these encoders. Suggested resale for 1 to 9 pieces at this writing is $345.

Litton Systems, Inc. (Encoder Division, 20745 Nordhoff Street, Chatsworth, CA 91311) offers up to 14 bits of resolution of BCD, binary, or Gray Code, and a solid state light source (LED) in a single-turn absolute encoder line.

Licon (a division of Illinois Tool Works, Inc., 6615 West Irving Park Road, Chicago, IL 60634) offers a unique noncontacting approach in their Series 33 position sensors. Two ferrite tube inductors are wired in series at right angles to each other near a rotating magnet. The position of the magnet affects the inductance of each ferrite tube, but the total inductance remains constant. An ac signal is impressed across the pair; the output at the tap is an ac voltage proportional to the shaft's position.

Licon also has an IC (the Licon 80-330057) that provides the driving signals for the transducers, detects the ac voltage at the tap, and provides a dc voltage output proportional to the shaft position.

So far, we've only seen one price tag on any of these. They're all pretty expensive, though, especially when compared to variable resistors. Even an expensive potentiometer is less expensive than an inexpensive absolute encoder. And we'll see that digitizing their outputs is not at all difficult.

Vernitech offers a number of superior potentiometers, offering such refinements as infinite-resolution conductive plastic film resistance elements, ball bearings fore and aft, a precious metal pickoff rotor, and alloy wipers in both single-turn and multiple-turn versions.

Clarostat Manufacturing Company, Inc. (Dover, NH 03820) manufactures a number of high quality potentiometers—the most expensive of which is under $30.

Spectrol Electronics Corporation (17070 East Gale Avenue, P.O. Box 1220, City of Industry, CA 91745) offers quite a good value in their Model 157 conductive plastic single-turn (340° continuous rotation) potentiometers. These little beauties have sleeve bearings, a life rating of two million shaft rotations, ±2 percent linearity, and 0.1 percent or better output smoothness. In quantity, they're about $3.

Another source of high quality potentiometers designed for precision servo feedback applications is Maurey Instrument Corporation (4555 West 60th Street, Chicago, IL 60629). Write them for specific specifications and price information.

Oops, almost forgot a significant source of digital absolute encoders—Baldwin Electronics, Inc. (BEI Electronics Inc., Digital Products Division and Industrial Encoder Division, 1101 McAlmont Street, P.O. Box 3838, Little Rock, AR 72203), who manufacture the Baldwin® line of encoders. Their 5VL80 series single-LED absolute encoders offer from eight to ten bits of resolution with several code options. Having overlooked them earlier, they are mentioned here because these encoders are often available in used but serviceable condition from a number of surplus outlets and MRO industrial equipment maintenance, repair, and refurbishing operations.

But the best bet for our android looks like the good old variable resistor style potentiometer. Part of the reason for this selection is a new family of A/D converter ICs from National Semiconductor Corporation that makes the whole project not only feasible, but attractive.

This family of ICs includes a number of inexpensive monolithic products offering 8-bit binary outputs based on a simple ratiometric input—meaning the three contacts of a pot. Several ICs are included—the ACD0800, ADC0801, ADC0808, ADC0809, and more. Our discussion here, though, will center around one of the most impressive members of this family, the ADC0816 Single Chip Data Acquisition System, a 40-pin CMOS IC capable of digitizing any selected one of sixteen multiplexed analog input channels under direct microprocessor control.

This is a 5-volt IC but can accept up to 15-volt logic levels at its control (digital) inputs. It has been expressly designed for use as a complete data acquisition system on a chip for ratiometric systems—systems in which the measurement is a percentage of some full-scale value and not necessarily related to any absolute standard. A potentiometer used as a position sensor is an excellent example: the voltage at the wiper is a fraction of the full-scale voltage across the pot and is proportional to its position. Unlike the digital control inputs, the analog inputs are only rated for a maximum of 5 volts.

Perfect! Take the regulated 5-VDC supply line that powers the chip and run one side of each of 16 potentiometers to it. Ground the opposite side of all 16. The 16 wipers provide the input signals to the chip. How's that for minimum parts count? National's data package includes specific instructions for interfacing the ADC0816 to the 8080, 8085, Z-80, SC/MP, and 6800 series microprocessors—but the IC can be used with any microprocessor, or even without one.

So now we see that standard potentiometers can be used simply, inexpensively, and conveniently to provide either analog or digital feedback on motor shaft position.

Before we move on from the subject of motor position feedback, let's spend a moment considering backups and failsafes. One important aspect of this is what happens at the end of travel for a rotating or levering joint. In an android, wires can tear, metal can tear, motor mounting hardware can tear, and there'll be some wear and tear on your nerves, too.

It's good design practice to include some headroom in case the inevitability of Murphy's Law hasn't changed. In the case of a 1½-revolution limitation on rotation, for example, it's a good idea to design hardware that can tolerate 2 revolutions or a tad more. And in the quite likely case that isn't precaution enough, some provision should be made for electronics that say "oops" before the mechanics say "ouch."

The case in point is end-of-travel sensing. Following our own advice, it should come into play before the end of travel is actually reached, but just inside or just outside of the ends of the anticipated (designed-for) control range.

Cams and microswitches have been used as limit sensors since my Dad was a boy. Interrupted LED/photosensors are often used these days. But an even less critical, highly dependable option exists in magnets and Hall-effect switches.

Texas Instruments, Inc. (Dallas, TX 75222) offers type TL170C, a silicon Hall-effect switch in a plastic 3-pin TO-226 package (meaning it looks a lot like a flat-front signal transistor). The TL170C incorporates built-in hysteresis (memory), which causes it to switch on when one pole of a magnet swings near it and stay on until either the opposite pole swinging by turns it off or power is removed.

One way to use these little marvels is to rotate a magnet that is oriented such that the switch is either on or off (your decision) in either a valid or invalid (again, your decision) locus-arc of rotation. Another way is to chop the supply (alternately turn on and off—in other words, use a square wave), in which case the proper magnet pole passing the switch provides the output. Either multiple magnets or multiple devices can be used to precisely define end points.

TI also offers type TL172C, a variation that is normally off, and have announced a linear device. Consult them for current information.

Micro Switch (a Honeywell Division, Freeport, IL 61032) offers a comprehensive line of Hall-effect sensors. Write them for their *Solid State Sensors* catalog, which includes a plethora of applications information, and their *Handbook for Applying Solid State Hall Effect Sensors*.

One of the most interesting devices they offer is the 9SS linear output Hall-effect transducers, which provide a voltage output proportional to magnetic flux. It can be used in conjunction with a Schmitt trigger or comparator as a limit alarm, as well as in applications requiring more precise position sensing.

Hall-effect devices require magnets. Here are some places to write to for information about magnets if you aren't happy with those your junk box provides: Arnold Engineering (Marengo, IL 61052); General Tire and Rubber (Evansville, IN 47707); Hitachi Magnetics Corporation (Edmore, MI 48829); Indiana General (Valparaiso, IN 46383); Temgam Engineering, Inc. (Otsego, MI 49078); Crucible Magnetics Division, Colt Industries (Chicago, IL 60639); Xolox Corporation (Fort Wayne, IN 46804); 3M Company (St. Paul, MN 55101); Bovee Engineering Sales Company, Inc. (Wheaton, IL 61087); Ceramic Magnectics (Fairfield, NJ 07006); LNP Corporation (Malvern, PA 19355); and TDK Corporation of America (Chicago, IL 60659).

If you happen to include radio control models as one of your hobbies, you're probably aware that servos are widely used to control steering, climbing, banking, speed, landing gears, and more. And you're probably aware that the most modern way of transmitting control information to the myriad of servos is a technique called *digital proportional*. You may or may not be surprised to learn that this technique is based on the width of a pulse of a fixed repetition rate, which is precisely the scheme we looked at to provide speed control for the main drive motors.

Exar Integrated Systems, Inc. (750 Palomar Avenue, Sunnyvale, CA 94086) has introduced a monolithic servo controller, type XR-2266. This IC, which is under $5, was specifically designed for use with radio controlled model cars. It incorporates two servo controller channels—one intended to provide speed and direction control and the other to provide steering control—plus detectors in the steering and direction circuits intended to light turn signals and backup lights. Joel Silverman, Exar Manager of IC Applications, has prepared an excellent paper, "A Monolithic Pulse-Proportional Servo IC for Radio Control Applications," which was published in the *IEEE Transactions of Consumer Electronics*, Volume CE-25, August, 1979.

Ferranti Semiconductors (Ferranti Electric, Inc., 87 Modular Avenue, Commack, NY 11725) offers another under-$5 wonder, type ZN419CE "precision servo integrated circuit." One interesting characteristic of this IC is that it uses the motor back-emf to slow the motor down as it approachs its target. Overall, the on-chip circuitry translates the width of the input pulse(s) into control for the motor. Outside the IC, motor, and pot, only half a dozen caps, half as many resistors, and two transistors are needed for a complete pulse-width responsive control system. At this writing, suggested U.S. resale is $3.80 in unit quantities, $2.20 in thousands.

Signetics Corporation (a subsidiary of U.S. Phillips Corporation, P.O. Box 9052, 811 East Arques Avenue, Sunnyvale, CA 94086) offers several interesting ICs. The NE543 Servo Amplifier includes a pulse

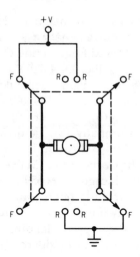

Fig. 12-1 *Example of "H" type motor drive using 4PDT switch. Actual circuits use transistors.*

width demodulator and motor drive transistors—up to 450-milliampere drive without external power transistors—on chip. It delivers a bidirectional bridge output operating from a single 3.6- to 6.0-VDC supply. Outside the motor and servo pot, just eight resistors and six capacitors complete the circuit.

The NE544 and NE644 Servo Amplifiers incorporate the above plus provisions for adjustable deadband and trigger thresholds. These ICs also incorporate a linear one-shot for high linearity (position versus pulse width)—0.5 percent or better.

Of course, we don't actually *need* any of these ICs. They're very nice, no mistake, but we can do better.

Consider the nature of the problem again. The potentiometer tells us where the motor shaft is, we know where we want it to be, and we know how to provide drive to the motor—all under microprocessor control.

Zang! There are no controlling pulse widths to generate except those for motor speed (and four bits—sixteen levels, eight in each direction—are plenty), and no complex circuitry is really required. A single-chip microprocessor with a PIA (or a micro that doesn't need a PIA—a Z8, for example), an ADC0816, and a handful of transistors (preferably VMOS FETs) can handle everything for *sixteen* motor-pot pairs.

The motors we're likely to use in such places as the wrist and fingers probably won't have to draw much more than ½ to 1½ amperes. Standard transistors and Darlingtons can readily handle these currents with not very significant losses. Motors can be driven with what is

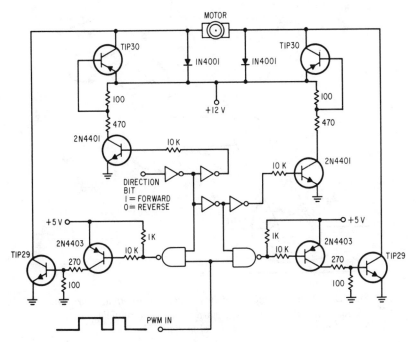

Fig. 12-2 *Bipolar transistor "H" type motor drive: one bit controls direction and logic controls PWM drive.*

variously referred to as a *bridge* or *H* configuration (see Figs. 12-1 and 12-2). Each motor lead is connected to the drive line with a pnp transistor and to ground with an npn. When the pnp on one side (lead) is on, the npn on that side is off and npn on the opposite side is on, and vice versa. So at any time, the pnp on one side and npn on the opposite side are on, their complements off. The result is an effective reversal of drive polarity to the motor without the use of relays. P-channel VMOS FETs can be substituted for the pnp's, n-channel for the npn's.

Also, the microprocessor can compare the current and target position values, determine error distance, and determine control motor speed to provide smooth movement without overshoot or hunting.

Aren't microprocessors wonderful?

There's one more aspect of this problem we want to consider before we move along. Imagine a simple elbow-style lever joint. Place a 1-pound weight at the end of the lever, then command the elbow to lift the weight a foot or so. What happens when the elbow reaches its target angle? The weight pulls the elbow open, of course, which causes the servo to seek "home" again. And unless a constant correction is applied, the arm looks like an elderly gentleman sprinkling sugar on his strawberries.

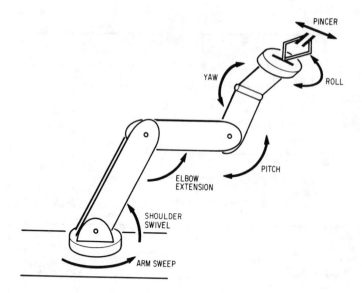

Fig. 12-3 *Model of industrial-type robot "manipulator" or "end effector" showing commonly accepted industrial terminology for various "degrees of freedom" (planes of motion).*

The motor does not have to be the source of that constant correction; instead, we could install a brake. The brake in this case would be a small solenoid-operated caliper-type arrangement that pinches a frictional disk attached to the motor shaft. The solenoid (probably with the aid of a spring) would hold the brake on with no power applied and release it just before a motor movement (by powering the solenoid), reapplying it when the motor has reached its target.

Now that we have thoroughly discussed the electronic aspects of these motors, let's take a look at the mechanics.

What motors? A more than fair question. Let's make a list.

First, we want to be able to tilt the trunk of the android 45° or so either forward or backward. Second, we want to be able to rotate the trunk 540° or so. Let's call these motions T/T (trunk tilt) and T/R (trunk rotary). We also want the head to be able to rotate 540° or so; call this H/R (head rotary).

The cameras in the android's visual system are going to have to be able to tilt through 270° or so; call this one C/T. Also, the binoptic system we'll describe provides parallax focusing by rotating two cameras in tandem over about a 90° arc. We'll see in our discussion of the visual system that a linear stepper motor and associated special circuitry provide a better answer to this particular need; nevertheless, let's assign a C/P (camera parallax) designation to this motor.

One option we'll want to permit the speech recognition system is an ability to locate a sound or speaker's bearing relative to the android head by rotating directional microphone ear shells. Call these H/R(L) and H/R(R), for hearing rotary left and right.

The several motorized joints in each arm closely parallel those in a human. Because some confusion is possible, the terminology used in industrial robots has been detailed in Fig. 12-3. (See also Fig. 12-4.) Hopefully, our references are going to be a little simpler.

The motor that lifts the arm (swings it forward or backward from the shoulder) at its top end we'll call S/L for shoulder lever. There will be an S/L(L) and an S/L(R), since we have both left and right arms—although with turn-and-a-half trunk and head rotation, it may get confusing when we try to identify which is which. (Just kidding.)

The shoulder junction is a box that can be lifted by S/L; this box also permits the arm to be rotated by S/R(L) and S/R(R), the left and right shoulder rotary motors. S/L and S/R can both be permitted about 540° of travel.

The shoulder box attaches to the "bicep" box, which in turn connects to the forearm box through E/L(L) and E/L(R), the elbow lever motor, which can be permitted approximately 300° of action.

The forearm box attaches to the wrist box, another 300° lever, W/L(L) and W/L(R) through a rotating joint permitted about 540° of rotation, W/R(L), and W/R(R).

Fig. 12-4 *Akron experimenter Andy Filo built what we see as an arm. This intelligent eye-to-hand motion perception system corresponds to the tongue of a frog catching a fly.*

Finally, at the far end of the wrist lever, is the box with fingers on the end. I suppose we might as well call it a hand, but it's really little more than a housing. Inside, packed around transducers and interfaces, are motors and mechanics to drive the three fingers and two thumbs of each hand. We will discuss this in greater detail later.

As you can see, we're dealing with a lot of motors and a fair amount of weight. It's important to recognize, again, that torque is by far the most significant design parameter we work with in designing the arms.

In the simplest case, for example, consider a 2-foot-long arm that weighs very little but has to lift 5 pounds. That requires 10 foot-pounds, or 120 inch-pounds, or 1,920 inch-ounces. But about the fastest you'd ever need to make that move is at 5 rpm—which means lifting the 5-pound weight through 30° in 1 second (or about a foot). That's only about a hundredth of a horsepower.

The Barber-Colman Company (Motor Division, 1300 Rock Street, Rockford, IL 61101) series CYQM-SZ 700 offers up to 150 inch-pounds continuous duty torque at 2½ rpm; with the load reduced to 6 inch-pounds, it can reach better than 100 rpm. TRW Globe Motors (an Electronic Components Division of TRW, Inc., 2275 Stanley Avenue, Dayton OH 45404) type BL motors (i.e., standard part number 100A108-5) with a mating planetary gear train (i.e., standard part number 102A205) offer the torque and speed range we've outlined with plenty of room at both ends. These are only two examples of literally thousands available from these and other motor manufacturers—and through surplus outlets.

The heftiest motor needed will probably be one of those at T/T, S/L(L), and S/L(R); next heftiest at E/L(L) and E/L(R). Examine carefully your lever-arms of torque, the masses and distances involved under maximum design load, and the speed of motion you want to permit. There's one heck of a difference between lifting a 20-pound bag of groceries and delivering a lightning-swift karate chop, but with proper design and selection, the same motor could be designed to accomplish both.

By the way, one of the inherent disadvantages of large-ratio gearmotors can belp us. Gears are a fairly lossy way of transmitting power, and that's true in both directions. Because of that, a minimal amount of braking at the motor rotor becomes a vast amount at the output shaft. In other words, we may not need an actual mechanical solenoid brake in some locations. Instead, some arrangement for shorting the motors should suffice.

At most, you now know how to arrange the motors and how to drive them. You still don't know where to drive them, or when. But, of course, there's still a substantial chunk of book left to read. Forge ahead and tallyho!

13
Taking the Heat Off

I remember the story of an amateur radio operator who had a problem with a transmitter output stage. No matter how well he tuned the thing, the losses and power dissipation were such that it always ran hot. His cabinet was well vented, and a fan forced air past the hot components, so everything operated reliably. But no matter what he would try, the cabinet developed a hot spot right over the output stage.

A few scalded fingers is all it takes to trigger typical amateur ingenuity. Our friend bolted a ceramic tile right over the hot spot and had a very unique rig as a result—the only one I've ever heard of with a built-in coffee warmer.

You *could* water cool your android, I guess, and tap off the coolant from time to time to make coffee. But it takes considerably less weight and energy to design the beast to run cool in the first place.

Consider the key heat-producing components within the android. These, of course, would be those devices that, for one reason or another, dissipate a great deal of energy. This might be because of large current-handling requirements, substantial internal losses, or a requirement (as in a regulator) to "throw away" power by converting it to heat.

We might note, one more time, that good design practice demands that each incidence of any of the above should be avoided, or at least minimized.

The largest current requirement on board the android—and thus the largest heat problem—is precipitated by the drive motors under start-up and stall conditions. Start-up is brief enough not to present a major problem, but stall currents may be prolonged. Stair climbing, for example, can require very high current levels for as long as it takes to reach the top of the stairway.

How long is that? Anything longer than several seconds can be a problem, in terms of heat dissipation. And we must limit the amount of heating the driving transistors experience. Bipolar transistors can destroy themselves through thermal runaway; at high temperatures, they conduct more current, which raises the temperature, which passes more current, which eventually destroys the transistor.

VMOS power FETs conduct *less* current as the temperature rises. While this mechanism eliminates the hazard of thermal runaway, it also reduces the current available to the motors at the precise moment when the motors require all the current they can get.

So, again, it is very much in our best interest to keep the temperature in the transistors from climbing too high. For a typical VMOS FET, for example, here's how the current relates to temperature (expressed, actually, as the percentage of full rated drive power versus the case temperature rise, in centigrade degrees, above an assumed 25°C room-temperature ambient):

0°	100%
10°	91%
20°	84%
30°	75%
40°	68%
50°	59%
60°	51%
70°	44%
80°	36%
90°	29%
100°	20%
110°	4%
115°	0%

These figures were determined by interpolation against power versus temperature derating curves on data sheet PD-9.303B published by International Rectifier for type IRF130, IRF131, IRF132, and IRF133 Hexfet™ VMOS power FET transistors.

The specific mechanisms in VMOS FETs that produce this reduction in available power include essentially a drain-to-source on-resistance that's proportional to temperature, which both increases the voltage drop across the device (reducing the voltage available to the motor) and reduces the current flow through the device.

For our example, let's take a look at what happens when each motor is driven by four IRF130 VMOS FETs, which are 12-ampere devices. You'll see in a moment why we might choose to incorporate four of these (for a total of 48 amperes) instead of two IRF150 (at 28 amperes each, for a total of 56 amperes). The IRF130 has a rated on-resistance of 0.18 ohms (specified maximum).

Twelve amperes through that junction produces a voltage drop of a bit over 2 VDC and a power dissipation of nearly 26 watts per device. If we had chosen the 28-ampere device with an on-resistance of 0.055 ohms, the voltage drop would be a bit over 1½ VDC, a power dissipation of just over 43 watts per device. Would you rather deal with 43 watts in two places per motor (total dissipation 86 watts) or with 26 watts in four places (total dissipation 104 watts)? And how significant is the 18-watt difference in terms of overall power requirements?

By now, it's fairly obvious that we have to pay careful attention to the selection of a heat sink. But whereas we might opt for the biggest available slab of finned metal in any other application, we ought to be more careful here. Heat sinks take space, add weight, and (if it matters) cost money. Watch closely, though: after we review the basics a bit, we can be just a little tricky and get by with some very neat solutions to the heat problem.

The job a heat sink has to do is to quickly conduct heat away from the device, then "share" it with the ambient air with as large a surface area as can be obtained. If the air in the immediate vicinity of the heat sink is still and confined, convection is the only way of accomplishing this exchange. *Natural convection* is the term used when the only air flow is that caused by the thermal differences betweeen the sink surface and the ambient air (disregarding infrared black body radiation, which here plays only a minor role); however, do not underestimate the power of natural convection, because this is the mechanism by which unbalanced solar heating of the Earth causes the common phenomenon of winds.

The reverse phenomenon also is of interest here. Just as uneven heating causes winds, winds cause an evening-out or leveling-off of temperatures between regions. On the smaller scale of our android, we can cause the wind to blow. The god we invoke is called a *fan*. If we find we have to use one, we can then use *forced convection*, which is simply convection in the presence of forced air.

This gives us two ways to better the performance of a heat sink: we can either increase its area or increase the flow of air across it.

The temperature rise of the device-plus-sink combination can be calculated as the product of the amount of power dissipated and the sum of three forms of *thermal resistance* (which is analogous to electrical resistance as a measure of resistance to the flow of heat): the three specific thermal resistances we're concerned with are the junction-to-case thermal resistance of the device, which is specified by the manufacturer, the thermal resistance from the case to the heat sink, which can be modified with the choice of mounting methods, and the thermal resistance of the heat sink to air. The usual terms for these values are degrees centigrade for temperature, watts for power, and °C/watt for thermal resistance.

Our task, in its simplest terms, is to shoot for the lowest thermal resistance values we can reasonably accomplish.

Looking at the two VMOS FET devices we've been discussing, the IRF130 is rated 1.67°C/watt, the IRF150 at 0.83°C/watt. With no heat sink, this would be the only thermal resistance factor, wouldn't it?

By now, you should recognize a trick question when you see one. Only the second term disappears, since there is no sink. The third term, thermal resistance between the sink (which is now the case alone) and air becomes significantly large. This value can be calculated as the reciprocal of the product of the surface area of the heat sink and a heat transfer coefficient, which depends both on the amount of convection air flow and on the specific units in which area is stated.

Obviously, the TO-3 case of the transistors offers only a minimal surface area, which results in a large value of thermal resistance.

Okay, so we need a heat sink, which means we need to mount the device to it. The normal procedure involves an insulating (electrically insulating, thermally conductive) separator between the device and the sink; it's normally coated with silicon grease or some similar substance in order to optimize thermal conductivity. These separators might be made of mica, fiberglass, or anodized aluminum, for example.

A much easier and highly effective substitute for the traditional approach is now available from Bergquist Company (5300 Edina Industrial Boulevard, Minneapolis, MN 55435) called Sil-Pads. These are available in shapes to fit virtually all standard diode, SCR, and transistor cases, or custom shapes can be ordered fairly inexpensively.

Sil-Pads 400™ are laminated silicon rubber and fiberglass, designed with pliable surfaces that assure good surface-to-surface contact for excellent heat transfer without silicon grease. Thermal resistance for standard 9-mil-thick Sil-Pads designed for use with TO-3 packages is within 2 percent of 0.50°C/watt; a newer, thinner 7-mil-thick Sil-Pad offers thermal resistance within 3 percent of 0.33°C/watt. This represents as much as a fifteen-times improvement over some traditional methods.

And now for the heat sink itself. There are many excellent sources for heat sinks, but we'll take a few shortcuts instead of the scenic route through all the options.

Aavid Engineering, Inc. (30 Cook Court, Laconia, NH 03246) is among those manufacturers offering a trendy new shape in sheet metal heat sinks, which might be described as a "quad bow tie" (see Fig. 13-1). Sides are folded up and out to make a compact but well-ventilated (and inexpensive) heat sink. Aavid's model for this design is series 5690; its thermal resistance is 5°C/watt.

In addition, Aavid offers a "booster" series 5791B, which attaches to the top of a TO-3 and provides some additional sinking, independent of the sink on which the device is mounted. This hat-like cooler demonstrates a thermal resistance of about 14.3°C/watt.

These two thermal resistances combine in parallel the same way (mathematically) that two resistors combine in parallel, yielding a net 3.7°C/watt for both devices used together. These figures are for natural convection only and will improve if forced air is introduced.

EG&G Wakefield Engineering (60 Audubon Road, Wakefield, MA 01880) series 690 heat sinks offer an almost identical design with the same approximate rating. The thermal resistance of these devices

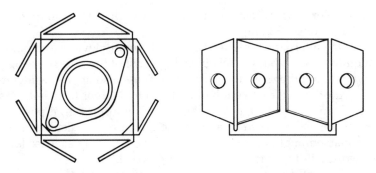

Fig. 13-1 *"Quad bow tie" heat sink configuration.*

is not quite linear, by the way, and may vary by about ±20 percent, but improve with increased demand (higher powers and higher temperature rises).

AHAMtor Heat Sinks (27901 Front Street, Rancho, CA 92390) also offers a similar model, series 490, which exhibits a slightly better thermal resistance, about 4.25°C/watt. Together with the Aavid booster, this brings our combined thermal resistance to about 3.3°C/watt. AHAMtor also was nice enough to provide price information on the 490: suggested U.S. resale at this writing is 56¢ for 1 to 49 pieces, 48¢ for 50 to 99 pieces, and 41¢ for 100 to 249 pieces. Imagine owning a hundred of these beauties for only about $40! It's an excellent investment.

Now we can combine the three series terms for thermal resistance and calculate the actual power versus temperature for our examples. This is 5.3°C/watt for the IRF130, 4.4°C/watt for the IRF150.

It's discovery time! Plugging in the device dissipation in watts for each device, we can see that the temperature rise for the IRF150 is 199°C and for the IRF130 is 137°C. Obviously, we are far better off using twice as many of the smaller-power device. Equally obvious is the fact that either temperature rise is intolerable.

Turning on a fan would indeed make a tremendous difference, limiting the temperature rise for the IRF130 to perhaps 68°C and for the IRF150 to perhaps 77°C. We have seen, though, that this much temperature rise limits the available power through the VMOS FETs to a full rated power of 40 to 45 percent. This reduction in power would bring the thermal rise for either device to about 31°C, which would increase available power to 75 percent of rated full power. Eventually, the cycles settle down and the IRF130 delivers approximately 17 watts at a 45°C heat rise; the IRF150 delivers approximately 26 watts at a heat rise of 47°C. "Delivers" in this sense refers to heat, not the load. Four of the smaller devices, as designed, are now driving 31 amperes; two of the larger ones, 33 amperes.

One alternative might be to parallel more devices and not drive them quite so hard, but the relatively high cost of these devices suggests that we find another solution. Space and mass also are a consideration.

Another alternative might be to go to a larger heat sink. Again, cost, size, and weight suggest seeking another approach, if possible. In any case, we will reserve this as an option in case our calculations prove unreasonable in light of the results of a breadboarding exercise later. In other words, if it's "back to the drawing board."

Perhaps a more expedient alternative is to take a closer look at the conditions under which our worst case is likely to occur.

Remember, we are using pulse-width drive to the motors. And remember that we want power (in the form of torque) and not speed at the triangular wheel drive during stair climbing, which is the worst-case power requirement we've determined. Since the discussions of heat in this chapter (and those in the literature, by and large) don't mention it, let's consider how quickly the thermal mechanism follows power flow.

Slow. Never mind lengthy calculations, just try frying an egg, turning off an incandescent lamp filament, etc, etc, etc. Heat is molecular motion—and it takes a while for heat energy to translate into and out of molecular motion, and a while longer to spread and distribute itself throughout a mass. Microwave ovens cook faster than conventional ovens because they introduce (or induce) this molecular motion almost uniformly throughout the mass being cooked. This motion increases only slowly, but because it is distributed throughout the mass of the food, cooking is accomplished rapidly. Conventional cooking techniques work more like a heat sink, spreading the heat slowly.

The result is (or is it the conclusion?) that we can consider time-averaged values for power requirements. And we can help that.

A thermal sensor at each group of devices can perform two functions. First, it can be used to turn on a fan when forced air is required. Second, it can be used to back off pulse width by providing temperature information to the controlling processor.

There. Now you know the options, know where to look for more information, and know a couple of tricks you can pull. That should take some of the heat off.

14
Collision Avoidance

You're in a very expensive car. It's nighttime. Your headlights aren't working, and you're alone on an unlighted stretch of highway. There's no moon and no stars. Even if the moon and stars were out, you couldn't see them because there's a blinding blizzard swirling powdery snow so quickly and thickly that you couldn't see the front of the car even if it were high noon at Yankee Stadium. You have no road map. The road is a tortuous one, twisting and turning, with trees and walls and cliffs and ditches and all sorts of bad things somewhere out there. And the insurance company notified you that they've canceled your coverage. How comfortable do you feel?

Now you have a pretty good feeling of what it's like for an android.

The right equipment in our expensive car could make up for the bleak picture the several obfuscations present. An infrared-sensitive camera could pick up heat images. Radar could locate objects around us. Navigational equipment could give us our bearings. And texture-sensitive infrared lasers could pierce the weather, bouncing back signals to identify what is road and what isn't.

So the problem for our android is not insoluble, at least. Now let's take a look at what some of the practical solutions might involve.

One tool available to us—one we will discuss in more detail elsewhere—is mapping. In short, mapping can provide the android with a memo pad record for each subdivided location in each space it enters. (This should all become more easily understandable in our look at mapping later.) Basically, the map for any given space can tell the android what kind of thing it can expect to find where and with what coarse level of probability.

Maps can be used equally well to locate where things are and where things aren't. They can be analyzed for best path without expending a great deal of exploratory energy. They can be experientially updated, and they can provide a heuristic programmer a number of "hints" for any number of purposes.

We're mentioning maps briefly here because maps provide a significant source of collision-avoidance data, and collision-avoidance

131

sensory systems provide a significant source of map data. While we may at times seem to be collecting more or less data than we need for either one of the functions, our goal is to collect just enough to support both.

For collision avoidance, the first thing we have to know is whether or not there's something "there." The second thing we need to know is *where*—what bearing and range. The third type of information—one we may or may not ever get ahold of—is *what* is up or out there.

In my younger and bolder (and perhaps more foolhardy) days, I recall driving the wee hours of a New Year's morning through the terrors of one of that year's worst zero-visibility blizzards, in the heart of the blizzard belt on a journey from Cleveland, Ohio, to Westport, Connecticut. I wanted to keep going because I didn't have enough money on hand to be able to afford a stop, but I wasn't anxious to kill myself. These was little or no traffic on the road, and mounting snows obscured any visual identification of where exactly the edge of the road might be. So I slowed to what I considered a prudent pace and pointed the car in as close to a straight line as I could manage. Inevitably, the car would leave one side of the road or the other. I became a human servo, steering slowly to get the car back onto the road and using the climb-on angle as an approximation of which way the road was really pointing under it all. The trip ended, eventually, without incident (although it may have been a few days before my teeth could unclench), but it did leave an indelible impression.

Many of the robot projects I've read about take the approach that my Dodge did that night—or worse. They provide simple microswitch bumpers, and only warn of a collision that has already happened. Don't get me wrong, I *recommend* installing this type of sensor; I just feel that these should be the *least* used switches on the android.

Noncontact collision avoidance (not collision sensing) is our goal. This requires a much more sophisticated level of circuitry, but the advantages are much like the difference between blindness and vision, between a three-dimensional video system and a white cane.

The first part of our collision-avoidance circuitry will be essentially optical in nature. A very narrow beam of infrared light is aimed along a well-specified path while a photodectector aimed and focused along the same path watches for any reflected energy.

For this type of system, only a small amount of the radiated energy may ever be reflected. There is no guarantee that the illuminated surface will be perpendicular to the light beam path. There's no telling what the surface will be either, from the flattest of flat black velvet to highly polished mirrors or reflectors to anything in between.

And there's no guarantee of signal-to-noise ratio, either. The noise, in this case, might be ambient light or rapid target motion, as the blades of a fan, for example, might present. And you don't want your android walking into the blades of a fan. Windows and glass doors represent another questionable return situation and one an optical system alone might or might not ever be able to resolve.

For optimum performance over reasonably good ranges of distance, incident angle, and surface reflectivity, it is of course in our best interest to provide both as bright a light source and as sensitive a detector as we possibly can. Let's deal with these two criteria before we discuss some of the other twists available to us—then again when we see how the twists change our requirements.

You might first think (as I confess I did) that our requirements for a pencil-thin beam and a very bright light make the diode laser a natural choice. There were objections, as with anything. Lasers are illegal in portable devices in some countries (like England, for one). And driving a diode laser requires fast, high-energy pulsing circuits—not all that hard to build, but something I don't have a lot of experience with or confidence in yet. I chatted with an author friend, Forrest M. Mims III, who has had a lot of experience with these circuits. Forrest gave me a number of important tips, including one that convinced me that lasers weren't the best way to go.

Here's the scuttle. Because of the substantial number of locations on and around the android where we're going to need a "something's there" sensor, we can't spend extravagantly. And extravagant expenditures are a requirement for lasers capable of continuous operation at room temperature.

The reasonably priced diode lasers that are available are *single heterostructure* (or SH) lasers.

All lasers require some minimum amount of current before they'll threshold into the lasing stimulated emission mode. For SH lasers, this is on the order of 5 to 10 amperes. Forrest correctly pointed out in his "Experimenter's Corner" on laser diodes (*Popular Electronics,* September 1977) that if you apply this much current to a laser diode it'll explode.

For SH lasers, there are two options—either liquid nitrogen cooling or a maximum pulse width of 200 nanoseconds. For example, RCA Solid State Division (Electro Optics and Devices, Solid State Emitters, New Holland Avenue, Lancaster PA 17604), offers ten devices in their SG2000 series of gallium arsenide injection lasers for pulse operation at room temperature. These devices offer total *peak* radiant flux power of from 1 to 20 watts at from 10 to 100 amperes peak forward current. (These ranges are over the field of ten devices; no one device exhibits this flexible a power characteristic.) The maximum permissible width of the current pulse is 200 nanoseconds, with a minimum of 200 microseconds between pulses—which sets 4,995 hertz as the maximum pulse repetition rate. This 0.1 percent duty cycle (maximum) limits the *average* radiant power to under 100 milliwatts, even under optimum conditions. And the narrow pulse width makes it a requirement that the detector be capable of very fast response; high-speed response and high sensitivity are hard to find in a single detector device.

We can attain much higher average radiant power with a LED, believe it or not. Forrest offered a High Current LED Pulser as one of his "Project of the Month" series from *Popular Electronics* (which isn't applicable here, alas). Forrest offers useful information here.

At 100 milliamperes drive, most high-quality gallium arsenide silicon (GaAs:Si) LEDs—which are infrared LEDs (or IRED, for infrared emitting diode), and we'll soon review the many good reasons for choosing infrared—emit on the order of 6 to 10 milliwatts of radiant power. This is comparable to the visible light a one- or two-cell penlight radiates—the kind with the little screw-in bulb with the lens on the end—which isn't exactly easy to see in a well-lighted room.

But the thing that limits the amount of current a LED can deliver is heat. Too much current for too long a time overheats the junction, murdering the LED. But if we limit the time during which it's applied, we can apply something like 20 to 50 times the continuous-duty current to the LED, resulting in a great deal more optical output.

An IRED rated for 100 milliamperes continuous forward current, for example, might also allow 10 amperes in a 1-microsecond-wide pulse repeated at up to, maximum, 200 hertz (a 0.02 percent duty cycle) and several watts of optical output. The same device also could permit other combinations of current, pulse width, and pulse repetition rate: for example, 10 microseconds at 7,000 hertz for 1 ampere (7 percent duty cycle, about 1/4 watt); 10 microseconds at 1,500 hertz for 3 amperes (0.5 percent duty cycle, about a watt); or 100 microseconds at 300 hertz for 2 amperes (3 percent duty cycle and about half a watt).

Don't take these specific numbers as gospel just yet. There are a number of other considerations that will influence the way we think of radiated optical power.

These figures have been heavily fudged from the published specifications of the type 1N6266 infrared emitter from General Electric (Semiconductor Products Department, West Genesee Street, Auburn, NY 13021). Other GE devices worth looking at include the F5D1 (880-nanometer wavelength), which includes a lens, and the F5E1, which has a flat window.

While you're investigating, take a good look at the TIES16 series of gallium arsenide IREDs from Texas Instruments, Inc. (6000 Denton Drive, P.O. Box 5012, Mail Stop 366, Dallas, TX 75222). The TIES16C, for example, is rated for a maximum of 3 amperes *continuous* forward current, producing minimally 350 and typically 400 milliwatts of radiant power output. In case your parts source or data library is as old as mine, this same device was formerly identified by part number TIXL16C.

Among Fairchild Optoelectronics devices from Fairchild Camera and Instrument Corporation (464 Ellis Street, Mountain View, CA 94042), the FPE520 seems most promising.

The SE-5455-4 from Spectronics (A Division of Honeywell, Inc., 830 East Arapaho Road, Richardson, TX 75081) looks like a good choice in their line. The most powerful device they offer is their SE-6478-2 (300 milliwatts continuous output at 3 amperes drive), which is available on special order only. Next one down is the SE-6477-060, rated 60 milliwatts continuous power out at 1 ampere drive.

General Sensors (A Division of Micropac Industries, Inc., 725 East Walnut Street, Garland, TX 75040) type GS 5040-3 is rated 6 milliwatts (typical) at 100 milliamperes drive.

The Xciton Corporation (Shaker Park, 5 Hemlock Street, Latham, NY 12110) type XC-55-PC also is rated 6 milliwatts at 100 milliamperes. Xciton's specification includes a maximum 10-ampere pulse operation rating for a 1-microsecond, 300-pps pulse train.

These LEDs can be driven by 555-type timers in their monostable mode to determine pulse width and by VMOS FETs to provide the necessary current plus high-speed switching with a very low component count. A current-limiting power resistor is recommended to assure that the LEDs' maximum current rating is not violated. A Signetics (a subsidiary of U.S. Philips Corporation, P.O. Box 9052, 811 East Arques Avenue, Sunnyvale, CA 94086) NE558 quad timer provides four monostables in a single 16-pin package. A single resistor and a single capacitor for each timer plus a pull-up resistor at each output are the only external components required. The monostables will trigger with a falling edge at the trigger input—no capacitor is necessary.

We'll see in a second just what it is that triggers the monostable, but first let's look for as simple an answer as possible to squeezing the last possible photon of usefulness out of the LED. Since we've already looked at getting the most out of it electrically, it's time to look at what improvements we can make optically.

The whole idea of improving the characteristics of the emitter is to make it more legible at the detector—more readable, that is. And there are a number of considerations that can lessen its readability—let alone the problem with nonideal reflectance at the variety of surfaces the beam is likely to encounter.

Signal-to-noise ratios are one significant problem area. It would be considerably easier to detect a dim light in an absolutely dark room, but as soon as it becomes a light among lights, it gets harder and harder to recognize.

One way we can make the task a little easier is to select a very narrow, well-defined slice of spectrum to both transmit and receive on. While the usual parameter used to define this selection is wavelength (usually in nanometers, occasionally in angstroms), we might just as well go by frequency or color—except how can you define a color in the infrared spectrum?

Infrared wavelengths are usually selected not so much because there are no other sources of energy there (there are dozens, all very powerful—over 80 percent of the output of an incandescent lamp is in the infrared, for example) or because we can't see them (a pity, but what's the difference?), but because the maximum sensitivity of silicon-based photodetectors falls in this portion of the spectrum.

By reducing the emissions and sensitivity of an opto system to a narrowly defined slice of spectrum, of course, we can simultaneously design it to reject other wavelengths. Even though a room is flooded with light, that light is usually a conglomerate of wavelengths, a menagerie of frequencies, a swamp of colors. The power-bandwidth of ambient light within our selected slice will be a small fraction of the total, while it can be as much as all of what our emitters put out.

So first we might want to consider an infrared transmissive filter in the path, preferably at the receiver. None is required at the emitter since IRED emitters are substantially monochromatic. Check the local camera store for particulars on Wratten filters for infrared—or read on for another solution. Still another source for IR products, by the way, is Edmund Scientific (101 East Gloucester Pike, Barrington, NJ 08007).

When you look through the Edmund catalog, don't rush past the simple lenses. You'll need a bunch of them to concentrate the IRED light into a beam and to concentrate the reflected light onto the detector.

Now, many materials from which lenses are built are excellent for transmitting visible wavelengths but are not very good at transmitting infrared. This is just a precaution, since we'll be working in the near-infrared range, with wavelengths just slightly longer than deep visible red. Most lenses you're likely to come across should work fine, but if the detector seems to work better without the lens than with it, you've found the gremlin.

Basically, the larger the lens diameter, the more light it can collect; the more severe its convex curvature, the shorter its focal length and the greater its transmission loss. Also, since infrared wavelengths are longer than visible wavelengths, the infrared focal length for a lens is longer than its visible focal length.

You can quickly estimate the focal length of a lens by focusing a distant light's image on a white card. The focal point is midway between the center of the lens and the image. For an even better approximation, focus the image into an intense spot by moving the lens closer to the card. The shorter the focal length of the lens, the more critical this positioning.

Okay. We've reduced the bandwidth and the beamwidth. There's still a trick available to us—and, imagine, we haven't even looked at detectors yet!

Richard Oliver of Centralab Electronics suggested a way to "Improve Photo Sensors with a Phase-Locked Loop IC" in the April 5, 1976, issue of *EDN*. Mr. Oliver notes that not even limiting sensitivity to infrared is an effective enough solution to the problem of ambient light problems for light emitter/sensor modules and suggests the use of an NE567 phase-locked loop tone decoder IC. The triangle-wave oscillator output (pin 6) of the device is squared and inverted with an op amp, which drives the LED (here, through an npn switching transistor). The phototransistor (in this circuit—other photodetectors could be chosen) couples the detected light signal to the input of the PLL (pin 3) through a coupling capacitor. The capacitor is important because it passes ac only and blocks dc, providing an input signal only for rapidly changing light levels (hopefully, the pulsed output of the LED) and not for steady-state ambient. The output of the PLL (pin 8) turns on when the transmitted energy is detected. Since the LED is pulsed by the PLL's own clock, phase lock is implicit—providing the transmitted light is *in phase* and not inverted one time too many.

A more thorough look at the 567 itself can help us optimize this circuit; a number of refinements and improvements in both technology and applications information for this and the other components permit us some room for improvement.

One note is that the 567 will lock on odd multiples of the center frequency; alternate odd multiples (5, 9, 13, etc.) will cause an output. A low-pass filter on the signal input line may be needed if there's any chance of adjacent modules reading each other.

The bandwidth of the PLL can be made very narrow by making the low-pass filter capacitor (pin 2) very large. At the frequencies we'll be dealing with, something in the range of 10 to 100 microfarads should provide a narrow lock bandwidth of about 1 to 2 percent.

The bandwidth of the PLL also is affected by the value of the output filter capacitor (pin 1), which also can be used to program a delay time before the PLL locks. Typically, this is twice the capacitance of the capacitor at pin 2—but that can lead to intolerably long delays.

There are two solutions easily available. One is the insertion of a narrowband filter in a unity-gain preamplifier stage between the photodetector and the PLL input. The other is to use a resistor to reduce the loop gain of the PLL. This is described in the manufacturers' literature. Write Signetics Corporation (a subsidiary of U.S. Philips Corporation, P.O. Box 9052, 811 East Arques Avenue, Sunnyvale, CA

94086) for literature describing their NE/SE567. The Signetics *Analog Data Manual* currently is available with a suggested U.S. resale price of $7.00; its companion *Analog Applications Manual* is $5.95.

Exar Integrated Systems, Inc. (750 Palomar Avenue, P.O. Box 62229, Sunnyvale, CA 94088) offers both an equivalent, the XR-567 Monolithic Tone Decoder, and a newer version that requires only about a tenth the power, the XR-L567 Micropower Tone Decoder.

The reason for our concern about narrow detection bandwidth is that we must operate a great many of these optical loops at once, and there's only minimal control over which might see which other at any time. The frequency range available to us is limited by the speed of the detectors we use and the power duty cycle requirements of the emitters.

With 64 such optical loops (not an entirely random choice), we can hardly afford to operate the PLLs at their 14 percent wideband (higher speed recognition) bandwidth. We can only fit about nine channels between 1 and 3 kilohertz (at 1,000, 1,150, 1,300, 1,450, 1,650, 1,850, 2,100, 2,350, and 2,600 hertz, for example; it would lock onto 3,000 hertz, which is 2N + 1 times 1,000 hertz), and we'll need more.

At a 4 percent bandwidth, we can fit about 21 channels, but lockup time is between 45 and 150 cycles (about 150 milliseconds, worst case). At a 2 percent bandwidth, we can fit (with a very tight squeeze) 54 channels—which should be enough, since we only have to prevent ten loops from seeing each other—at 100 to 300 cycles lockup delay, which is between 33⅓ and 300 milliseconds. Not even counting the time it takes for the processors and controls to respond to the sensors, the android would travel nearly a yard before it even recognizes that something is there.

Enough mystery. We're going to settle on a 3 percent bandwidth, using a minimum capacitance in the 567's low-pass filter but using a resistor to reduce loop gain. This results in a maximum of about 70 cycles before lockup. We can increase the band of frequencies we're working with to roughly 2 to 10 kilohertz. And we can be *very* careful about which frequency we put where.

Let's take a quick look at those "wheres." Sixteen of these optical loops ring the top of the trunk cylinder in one approach to radial sensing. If nothing else, these will locate the greatest percentage of possible obstacles most of the time.

The front of the trunk cylinder is a good place to put a vertical column of eight. These will help identify the height of an obstacle.

Four at the front and four at the rear bottom corner of the carriage chassis can locate ground-level obstacles. Four small turrets (one on each of the four corners of the subchassis) might contain another four optical loops each, aimed at 45° increments from straight out front/

back through straight out sideways to overlapping side coverage. These provide full "whisker" coverage to help the android know whether or not doorways are wide enough, whether it's clear for a turn, and so on. This medium-low coverage complements the trunk-top ring for full radial coverage. Also, the same four frequencies can be used in each turret, repeating clockwise in order through the 135° arc, without any one loop seeing any other.

Similarly, the frequencies used in the eight optical loops in the vertical column can be used again in the trunk-top ring—on the opposite side (I guess we might call it the back, but that's arbitrary). Also, the four ground-level loops in the front can use the same frequencies as those in the back.

So far, we've only located 48 optical loops. Whether these are enough, are too many, or more are needed is a decision to make yourself once your beastie is a little better defined.

But we've left our loop open, so to speak. Let's get back to discussing the circuitry . . . noting that the 48 optical loops we've identified so far require only 24 different frequencies.

The application literature suggests a number of ways of increasing speed while reducing bandwidth. One is to increase amplifier input sensitivity (pin 1) by placing a resistor in parallel with the output filter capacitor (to ground). This reduces turn-on time, but not turn-off time (which is less important for our application). Another is to use the output (pin 8) to switch a second, large-value bandwidth-reducing capacitor in parallel with the initial output filter (capacitance in microfarads is 130 ÷ frequency for minimum, 10-cycle lockup) using a pnp transistor. This means an initial larger bandwidth, which is immediately constricted once anything appears in band. A third approach in the literature is to apply a voltage to pin 2 through a resistor. A 270-ohm resistor from pin 2 to the junction of two 470-ohm resistors (which are connected to the plus and ground lines, respectively) will limit the bandwidth to about 2 percent, at the expense of a larger capacitor value to ground for a given cutoff frequency.

And around and around we go. This is a good time to suggest trial and error as the best way to narrow these design loops down quickly and specifically for the particular devices you use. Solderless breadboards are my personal favorite way of trial-and-error fine tuning. Contact Global Specialties Corporation (70 Fulton Terrace, P.O. Box 1942, New Haven, CT 06509) for their catalog and information about what is probably the most complete line of solderless breadboarding products in the world. They also offer a system (The Experimentor System™) for quickly translating breadboards into printed circuit boards using a breadboard-modeled prepunched and pre-etched PCB.

Which only leaves the choice of a detector. We'll both widen and narrow this quickly.

First, silicon infrared detector technology has been improving steadily. There are literally dozens of sources for the devices we're talking about. The limiting factors are speed and sensitivity, and they're very hard to find in the same device. This is understandable if you consider that to achieve high speed, there isn't much time for photons to be detected; to achieve high sensitivity, photons may need to be accumulated over some small chunk of time.

For this reason, standard phototransistors and photodarlingtons are *not* the way to go. Photodiodes and FETs offer two better choices.

A photodiode may be connected across the input of an operational amplifier, anode to plus, cathode to minus. A 5-megohm resistor from the plus input to ground, another 5-megohm resistor from the output to the minus input (feedback), and a 100-picofarad capacitor from the compensation pin to ground completes the circuit with an LM108 or equivalent.

There is currently one single best choice for this photodiode: the BPW41 silicon P.I.N. infrared detection photodiode from Ferranti Electric, Inc. (Semiconductor Products, 87 Modular Avenue, Commack, NY 11725). It features a low junction capacitance for fast response, a 7½-square-millimeter active area for good sensitivity—*plus* a *built-in* infrared transmissive filter at 925 nanometers (peak, 50 percent from 730 to 1,040 nanometers), meaning virtual immunity to visible radiation. At this writing, suggested U.S. resale is $2.88 for 1 to 24 pieces; $2.42 for 25 to 99 pieces; and $2.02 for 100 and up. (Notice that 84 pieces cost more than 100 pieces, and 25 pieces cost you only 2$^{¢}$ more than 21, so don't be bashful about ordering a few more than you think you need.)

The other option worth considering is the FOTOFET®, like the FF108, for example. Teledyne Crystalonics (147 Sherman Street, Cambridge, MA 02140) manufactures these photosensitive FET transistors, which offer the high speed and high sensitivity of a FET with low dark current in a windowed or lensed package.

I've resisted the urge to give you suggested schematics for optical loop circuitry, knowing that only by developing your own will you meet the specific requirements of the needs of your android and your available parts—from filters to lenses to emitters and from detectors to drivers to support circuitry. It's worth the time, though, because of the large number of loops required. Without attention to detail, optical loops could end up eating a $2,500 chunk out of your casual cash. With a little care, you could cut this by a factor of ten or more. Enough said?

These optical loops fulfill the role of the "something's there" sensors we talked about—and enough of them can locate that "something" radially. Now we need to look at how we might be able to locate it more absolutely.

The best approach is to follow the lead of the people who have spent considerable time and money researching this requirement for needs of their own—Polaroid Corporation (Ultrasonic Ranging Marketing, 20 Ames Street, Cambridge, MA 02139). The ultrasonic rangefinder apparatus used in Polaroid cameras is one of the most sophisticated ever developed.

A 420-kilohertz quartz crystal oscillator is clocked into a four-frequency "chirp" a millisecond long. The chirp is amplified and outputted through a sophisticated electrostatic transducer. The receiver waits 600 nanoseconds for the vibrations on the transducer's diaphragm to decay and then starts listening for the echo. Ramp gain increases the receiver sensitivity with time, since a longer echo means a weaker return signal.

Polaroid has announced an Ultrasonic Ranging Designer's Kit. It includes a technical manual, two transducers, a modified electronics circuit board, a circular polarizer, two batteries, a battery holder, and assorted hardware. Price of the kit at this writing is $125, subject to change and local taxes. The material lets you build a rangefinder that displays five readings a second within a range from 0.9 to 35 feet.

If that price doesn't appeal to you, perhaps you'd be interested in rolling you own. It isn't terribly hard, thanks to the LM1812 Ultrasonic Transceiver from National Semiconductor Corporation (2900 Semiconductor Drive, Santa Clara, CA 95051). Originally designed for use as a "fishfinder" with water-coupled transducers, this IC includes a basic SONAR system, requiring only a few external components. These components do require a certain level of explanation and understanding, though—so, shall we proceed?

We'll be designing our SODAR (the air-coupled version of SONAR) for operation at 40 kilohertz, which is adequately beyond the hearing range of both humans and animals.

These transducers can be used for both transmitting and receiving, offering 0.2 watts cw (continuous wave) transmitting power (maximum) and -60 decibel (reference 1 volt per microbar) receiving sensitivity. They are completely sealed, aluminum-housed units, making them virtually impervious to dirt, dust, smoke, and water damage. Transducer capacitance is 1,600 picofarads, and tuning inductance is 10 millihenrys, typically, and price is $10 or less.

The LM1812 output impedance is on the order of 10 ohms. National's Tom Frederiksen tells me that one of the most common reasons problems are reported with the LM1812 in SODAR applications is insufficient magnetizing inductance in the primary of the output autotransformer. If the inductive impedance of the primary is only the 10 ohms (minimum) required to keep the IC output from blowing itself out, that means 1.2 amperes as 12 volts at the output—14.4 watts—71 times the cw rating of the transducer.

Okay, inductive impedance equals inductive reactance equals 2π times the frequency in hertz times the inductance in henrys. Power equals the voltage squared divided by the impedance. We know the frequency is 40,000 hertz. Let's assume the maximum power we can tolerate is 20 percent more than the transducer's cw rating, or 0.24 watts, and that this power level occurs at the full-charge voltage, about 14 volts. This means we're looking for about 817 ohms. Plugging that into the inductive reactance formula, we find that we're looking for a minimum 3.25-millihenry inductance.

Note that we earlier saw the tuning inductance required in the secondary is about 10 millihenrys. Let's fudge a bit and specify a 3:1 ratio—3.3 millihenrys in the primary, 9.9 millihenrys in the secondary.

As far as I know, this isn't a stock transformer anywhere. So let's look at where to go to get what you need.

Permag Corporation (88–06 Van Wyck Expressway, Jamaica, NY 11418) is a good source for ferrites, like toroids and cup cores, if you decide to wind your own. Permag also offers ferrites from Siemens Corporation (Components Group, 186 Wood Avenue South, Iselin, NJ 08830).

Another source is Magnetics, Inc. (Division of Spang Industries, Inc., P.O. Box 391, Butler, PA 16001). Write for Design Manual TWC-300R, catalog FC-208, and catalog MPP-3035.

Another source, especially noteworthy, is Renco Electronics, Inc. (240 Old Country Road, Hicksville, NY 11801). In addition to offering cores and magnetic components, they also can build custom transformers. Specify a 3.3-millihenry primary, a 9.9-millihenry secondary, and 40-kilohertz operation. Renco also manufactures an RF choke perfect for the tuned RF circuit on the LM1812. Renco part number RL-1284-4700 is a 4.7-millihenry choke with a Q (quality factor) of over 120—a minimum Q of 20 is required.

The other components required by the SODAR circuit are fairly straightforward and are clearly spelled out in National's LM1812 device literature; but that doesn't mean that a few more hints aren't called for.

First, the maximum range you decide to allow times the speed of sound in air (1,090 feet/second, 13,080 inches/second) dictates the minimum time between 1-millisecond-wide pulses to the transmit modulator.

Before we go any farther, let me suggest three instruments that can make the work a little easier. All three are available from Global Specialties Corporation (70 Fulton Terrace, P.O. Box 1942, New Haven CT 06509). The Model 3001 Digital Capacitance Meter is a precision $3\frac{1}{2}$-digit capacitance-measuring instrument covering a wide range, from a few picofarads to 0.2 farads. This is excellent not only for selecting capacitors precisely but also for measuring the capacitance of other components, like cable and transducers.

The Model 4001 Pulse Generator offers independent control over pulse width and pulse spacing, a number of output modes, and a number of output levels—including TTL (transistor-transistor logic) and high and low variable outputs. This lets you create just about any duty cycle pulse—singly, continuously, gated, or however you need it.

The third instrument is the Model 5001 Universal Counter Timer. This measures frequencies, periods, intervals, frequency ratio (which can be used to indicate duty cycle), and counts events.

Now let's take a look at the pin-by-pin circuit requirements. If you don't have your LM1812 data yet, return to this section later.

Pin 1 is where the tank circuit that is shared by the transmit and receive sections is connected—a 4.7- to 5.0-millihenry coil with a Q of 20 or better in parallel with a 3170-3370 (3300 nominal) capacitor to the positive supply line sets up a 40-kilohertz frequency—dead on target for the transducer.

Pin 2 is the input to the second RF stage. It connects to the wiper of a 5K gain pot through a 0.01-microfarad capacitor. The gain pot connects between pin 3 (first receiver RF stage output) and ground.

Pin 4 is the input to the first receiver RF stage. Since we're using the same transducer as both sender and receiver head, this pin connects through a 5,100-ohm resistor and a 510-picofarad capacitor to the junction of the top of the autotransformer secondary and the center connector of the coax feeding the transducer (or the transducer itself, if board mounted). If we were using separate send and receive transducers, we could connect the receive transducer to this pin through the capacitor alone, or we could insert a preamplifier in line.

Pin 5 is one of three grounds provided and corresponds to the transmitter power amplifier. Pin 6 is the output of that amplifier and connects to the tap of the autotransformer through a pulse-stretching circuit. An RC "T" network consisting of a 20-ohm series resistor into a 0.02-microfarad capacitor to ground and a 160-ohm series resistor feeds the pin 6 output to the base of a 2N4354 pnp transistor; a 510-ohm resistor to the positive supply biases the base and a 250-microfarad capacitor helps filter the current demands of this stage on the supply (it connects from plus to ground). The emitter is connected directly to plus, the collector directly to the autotransformer "tap"—the junction between the finish of the primary winding and the start of the secondary. The start of the primary is connected to ground, the finish of the secondary to the transducer.

Pin 7 is the output of the modulator stage, which is triggered on for 1 microsecond at the beginning of each measurement cycle. This is accomplished with a pulse approximately 1 millisecond wide applied at pin 8, which also switches the transmit/receive switch gate; the T/R switching times are shaped by the 0.47-microfarad capacitor from pin 9 to ground.

Pin 10 is the second of three grounds, this one at the receiver threshold detector. The 0.68 microfarad capacitor from pin 11 to ground determines the transmit/receive duty cycle.

Pin 12 is the positive supply connection to the transmitter power amplifier and the balance of this part of the circuit, at least; there also is a connection through the tank coil at pin 1.

Pin 13 is a transmitter power amplifier connection that is used in SONAR applications of the IC, but not in our SODAR; we'll leave the pin unconnected.

Pin 14 is the output of the display driver transistor (it's the collector of a Darlington pair). A 5,100-ohm resistor connects from here to the positive supply. A negative pulse (from the supply down to about a volt) appears at pin 14 and can be used to latch the output of a counter. The clock rate of the input to the counter (or more accurately, its period) determines the units of distance being counted.

Pin 15 is the third ground, this one at the display driver. Pin 16 is an auxiliary display control used with neon lamp "fishfinder" circuits, but not needed here. We will leave no connection at pin 16.

Pin 17 is the input of the integrator stage of the receiver, and we will connect a noise rejection filter and control here. First, a 1-microfarad capacitor connects from here to ground. Second, a 22K resistor in series with a 25K trimmer connects from this pin to the positive supply.

Pin 18 is part of a pulse train detector, and a 0.01-microfarad capacitor connects from here to ground. The LM1812 is designed so that before a valid output is recognized, a minimum number of returned (received) pulses must first be detected; if this pulse train is interrupted and as few as two or three are missing, the pulse train detector dumps the integrator. Together, this assures an output only for valid echoes and the rejection of spurious impulse noise.

The T/R switching circuit on board the LM1812 can be circumvented by applying a logic low at pin 9; this turns the receiver off. This permits us a rather nifty and headache-saving little trick.

We're going to want to use a number of SODARs on the android. Four will ring the trunk at 90° intervals, and the fifth will be in the camera tilt box on the head platform.

We can use microprocessor control to turn on our SODAR receivers one at a time, trigger the transmitter, time the return, store the result, turn off the receiver to prevent false results from subsequent echoes of the same return or from echoes of other SODARs, then move on to the next unit and repeat the cycle.

The LM1812 has been written up by Frederiksen and W.M. Howard as "A Single-Chip Monolithic Sonar System" in the *IEEE Journal of Solid State Circuits*, December 1974, Volume SC-9, Number 6. It also received some attention from Lou Garner in his July 1976 "Solid State" column "Build Your Own Sonar System" in *Popular Electronics*.

The last part of the collision-avoidance system we want to mention is the part that responds to the unavoidable collisions—the bumpers. A number of clever configurations are possible, most of which depend on the eventual actuation of some basic switch.

There are more kinds of these switches available in surplus stores alone than there are kinds of candy in a candy store. And there are nearly as many manufacturers.

Still, one variation I've seen has impressed me enough that I've chosen to share it with you. This is the Type D2MQ sub-subminiature basic switch from Omron Electronics, Inc. (Control Components Division, 650 Woodfield, Schaumburg, IL 60195). These microminiature marvels measure 8.2 by 2.7 by 10.6 millimeters and are rated 10,000 operations at 50 milliamperes. These are great not only for bumpers but for limit switches in places as small as fingers and as position sensors with cams.

That just about covers our notes on things that shouldn't go bump in the night. But keep the collision-avoidance problem in your thoughts as you read about the visual and mapping subsystems. With androids, you have to go to all kinds of trouble just to keep the beastie from getting into any himself. And that's a programming problem I still haven't solved for my two naughty cats.

15
Fingers

Many nights I've stayed awake wishing that "Thing," the disembodied hand from the old *Addams Family* TV series, was available commercially as an off-the-shelf product. How much that would simplify the design of what roboticists call "end effectors"—you and I know them as *fingers*!

Consider the problem of trying to design a flexible linear joint that hinges in several places, curls or straightens on command, has a rigid skeletal structure yet a flexible and pliable semifrictional outer surface. Then consider that this same device is rich with sensors. There's a sense of pressure, a sense of position, a sense of temperature, and more.

Human hands are constructed to favor grasping. While the upper two knuckles in each finger only bend about 90° and only in one direction, the lower knuckle adds about 30° of permissible, controllable motion at a right angle to this. The thumb is a swivel joint capable of moving through roughly a quarter hemisphere of space.

I've seen pictures of robots people have built, and not many of them come close to a tenth this capability. There are pincers and claws, grippers and paws, and just about every simple mechanism imaginable.

But I've found a way to come close to the action of actual fingers. I'm not alone in this discovery. There are a number of prosthetic devices, I later learned, which use much the same principle. There are toygrasping hands that use some of these principles. And the Japanese have developed a three-finger hand that does very much the same kind of thing—again, a discovery I made after many sleepless nights trying to find some simple, obvious answer to my needs.

What follows in this chapter is a description of an android hand based on three anthropomorphic fingers opposed to two similar thumbs (see Fig. 15-1)—a compromise necessary since I have yet to discover how to build my thumbs with a simple swivel joint (but I hope you have better luck). The fingers are equipped to sense pressure and can be equipped to sense temperature as well. And while this hand is rather oafish in its construction (alas, my android will never be a neurosurgeon), it permits the android to work with most common hand tools, maneuver glasses and bottles and kitchen utensils, turn doorknobs, push buttons, and more. Parts of the solution to the many design problems are surprisingly simple; others surprisingly available. And now that I've properly teased and intrigued you, let's get down to cases.

Let's start with a look at the human skeleton. The bones of a finger are rigid linear sections joined with one-dimensional (degree of freedom) hinges. I thought of bolting hinges together, copping watchbands, designing custom metal parts and more until I discovered a simple mechanical contrivance that fills the bill perfectly. It consists of rigid linear sections joined by simple hinges. The technical name of this device—which by now, I'm sure, you're dying to guess or know—is "chain."

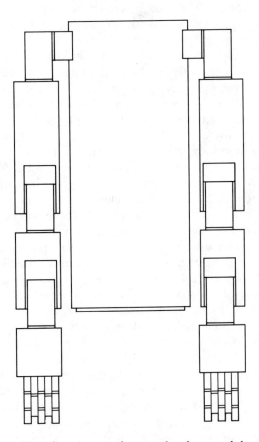

Fig. 15-1 *Three fingers opposed to two thumbs on each hand complete the android's human-like reach-around and grasping capabilities. Note also the modular approach to arm motor drive box design. Turntable bearings at each swivel permit wires to pass through.*

Gasp! Heavy sighs! Curious "ohs"! Thrills! Moans! Groans of disappointment! Bemused chuckles! I don't know what you're going through, but I went through them all. My wife thought I was reading the letters section of *Penthouse*.

Okay, big deal, we can use chains in place of bones, but that doesn't have anything to do with the muscles. How can we make the fingers dance to our commands, let alone the android's?

Taking another look at the muscle and bone model of the android hand at the end of our own arms, we find one set of muscles along the inside of the joints, another along the outside. As one set tightens, the other relaxes. These muscle pairs are common in the design of humans. They're called *flexor-tensor* or *flexor-extensor* pairs.

We can do the same trick with wires.

First, though, let's modify the chain a little to bring it a little closer to the way our bones work. While we could always design a double-jointed android, the farthest a finger usually extends is straight out. By welding a small stop plate at each link-hinge on the chain, we can keep the chain from flexing beyond straight out. As long as we're welding (chances are you can get a guy at a local gas station to do this for you for a few bucks), we can extend the distance between knuckles by welding a plate on both ends of a couple of link-hinges between those we want to use as knuckles. Depending on the pitch (link length) of the chain you get ahold of, you might want to allow two or three links at the fingertip before the first half-free hinge, another two or three before the next, and so on. It's also a good idea to allow two or three links beyond the bottom knuckle to give you an easy way to mount these fingers to the android's hand. There will be another bit of welding in a minute, so don't run out just yet.

The cable we use in the fingers is a crucial part of the mechanism and should be carefully chosen for strength and durability. You don't want to use twine or fishing line here or you could find your favorite bottle of 12-year-old Scotch hitting the floor when your android loses his grip.

Cable Manufacturing and Assembly Company (19 Gardner Road, Fairfield, NJ 07006) makes Cycle-Flex™ miniature cable and assemblies, manufactured from stainless steel with or without a nylon coating—a desirable plus to reduce friction and wear on components. These are available with eye fittings, loop fittings, and others. They offer something on the order of a million operating cycles of service life.

Sava Industries, Inc. (P.O. Box 150, Pompton Lakes, NJ 07442) makes Hi-Flex™ miniature and small stainless steel cables and assemblies, bare or with nylon, vinyl, or Teflon coatings. They also manufacture a number of types of pulleys and accessories you'll need.

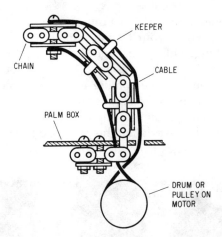

Fig. 15-2 *In this model of android finger mechanism, groups of chain links are held rigid with metal strips. Mounting screws also hold "keeper" guides for flexor/tensor cables, which are around motor drum or pulley.*

Each finger (see Fig. 15-2) requires either one or two cables, depending on how you anchor them on the motor end (more on this in a moment). Either way, two free cable ends are bolted or otherwise fastened to one of the links of chain past the top knuckle, near the fingertip.

Either midway between knuckles or on either side of each knuckle, you'll need to weld some sort of keeper in place to keep the wire close to the chain. This can be as simple as a small flat tube or loop which can be round, square, or any convenient shape—or as sophisticated as a tiny roller bearing perpendicular to the path of the wire and parallel to the hinge joints on the chain.

This approach to finger mechanics (shown in Fig. 15-2) permits several easy refinements through welding or the use of cements (to eliminate most of the metal strips and mounting hardware), which includes lower profile "keeper" loops, an angle metal mechanical tie to the palm box, a lightweight chain, and more.

When you have your first finger put together (no need to go through all these paces, you can rig up a dummy from strips of plastic with electrical tape as hinges, pipe cleaners as keepers, and string for the cables), clamp the bottom end of the thing in a vise. Pull slowly on one wire while maintaining pressure on the other. Behold! Thou hast created a finger!

Now the hard part is to keep the group of fingers from being all thumbs. For as long as you've heard the expression "opposed thumb" and assumed that it meant a thumb opposite your fingers, have you ever looked at where your thumb actually is? It's more of a sidewinder than anything. Still, your hand does things you want it to.

The fingers we've just designed are okay for curling, but it's tough to imagine one of them coming in from the side like a human thumb and getting anything accomplished.

But—aha!—your thumb *is* opposite your forefinger (or can be, anyway). Together, they work like the pincers on one of those gadgets stores use to grab packages from high shelves.

Fig. 15-3 *Low cost pressure sensors. National Semiconductor's monolithic LX0503A IC comes in a TO-5 package for easy printed circuit board mounting. (Courtesy of National Semiconductor)*

Okay, the anthropologists like the idea of an opposed thumb, so let's put our android's thumb opposite his fingers—really opposite them, where the middle of our palms would be. And like our thumbs, let's save some chain and only give it two knuckles instead of three. Good enough?

Nope, not quite. Try picking up a hammer with just your thumb and forefinger. Maybe you can control it, maybe not. Either way, it's just as easy to hedge our bets by departing from nature a little and placing two thumbs across from three fingers. Space them apart just a bit so the thumbs can fold with their tips between the curled fingers, and vice versa (since it wouldn't do much good to have them running into each other). Now that's a grip! And your android is well on its way to being able to handle just about anything you can.

Now to the problem of what in the android's hand is going to pull on one side of the cable while easing the tension on the other. Hopefully, the answer will be something we can work with a motor.

The simplest answer would be to use a small stepper motor and a simple pully or drum. Even a half-inch pulley would wrap enough cord in a single revolution (or half a one each way) to fully curl or extend the fingers. The next simplest answer would be to use a small, slow gearmotor, plus a pulley or drum. The next best answer would be to drive the pulley or drum with a not so slow motor, but drive it with a low duty cycle pulse. A simple pot provides position feedback in each case.

One thing we may not have mentioned even in our several discussions of motors is that a given motor driven by the same voltage, whether constant or pulsed, will draw varying amounts of current depending on the load on the motor. In the hand, for example, the fingers curling in free space draw less current than the fingers curling around the bulb of a bicycle horn, which draws less current than the fingers curling around a rock, and so on.

This motor current is easy to sample—a small resistor in series with the motor or its drive circuitry will develop a voltage across it proportional to the current through it. This momentous discovery was originally made by a certain Mr. Ohm some years ago.

Motor loading is one form that the sense of touch for an android might take, though a crude one. A much more elegant solution, quite reasonably priced, now exists in the form of a Monolithic Pressure Transducer in a TO-5 package, the LX-0503A (see Fig. 15-3) from National Semiconductor Corporation (2900 Semiconductor Drive, Santa Clara, CA 95051).

Interestingly, National's Art Zias presented a seminar on this device entitled *Finger Tips of the Robot*—an amusing pseudohistorical look at the growth of the transducer marketplace and less than flattering to the poor dupe robot models he talks about. To paraphrase a few portions:

> The language and training gap between the new technocrats, called Electronikers, and the old technicrats, called Mechanikers, leads to inbreeding, with two inevitable results. The first is the emergence of both unmechanikered computers and un-

computerized mechanics; the second is *anti*anthropomorphization and machine paranoia. Together, these lead to the "Helen Keller robot," which can do everything but has no sense of what it's doing—a relatively senseless (in sensory perspective) combination of computers plus activators.

Art's demonstration mechanism consists of some hand-operated pumps, solenoid valves, transducers, computers, and gas bottles. The nicknames applied to the apparatus were a little too off-color to repeat here, but the lecture's conclusion was "America loves seducers, producers, sex, and transducers from National Semiconductor!" I can't wait until it plays at the local drive-in.

Now, absolutely the first thing you're going to have to do is contact National for one of the best application manuals you'll ever read, their *Pressure Transducer Handbook*. While you're at it, ask for Application Note 218, "A Pressure Microcontroller: Pressure Controlled by Microprocessor with Digitally Interfaced Pressure Transducer and Solenoid Valves"; Application Note 234, "A Microprocessor-Controlled Pressure Regulator"; and the LX0503 Series Monolithic Pressure Transducer data sheet. Here's some of what you'll learn.

The LX0503 is an absolute pressure transducer with a tube inlet fitting on a TO-5 can that provides an output voltage proportional to the pressure on the membrane over the piezoresistive elements inside. With an LM324 and an LM331, you can build a simple single-supply circuit with a frequency output proportional to pressure from 0 to 30 psi.

Great! Only where does that pressure come from? Art Zias, Pressure Transducer Marketing Manager Ray Pitts, and the gang at National shared a few intriguing ideas with me.

With a minimum number of transducers—one per hand—we can get "squeeze" feedback by putting a toy balloon on the surface of the palm of the android's hand, between the three fingers and the two thumbs. To protect the balloon, we can cover it with a piece of rubber glove. The balloon could wrap around the pressure inlet tube on the LX0503, provided we never exceed 30 psi. And we're not likely to, or we'd be inflating our 28-psi auto tires by mouth. But if you're worried, invest in a tire pressure gauge. For something a little more durable than a toy balloon, contact the anesthesiologist at a local hospital and ask about where you might be able to buy a respirator, the rubber air bags that expand and contract as surgical patients breathe.

When you get to wherever he sends you (or to your corner pharmacy), ask about buying the squeeze bulbs at the ends of eyedroppers; don't be disappointed if you have to buy the pipette, bottle, and all—it won't be a budget crippler. These little squeeze bulbs are part of another scheme, one that places pressure sensing along each finger, right to the tip. The squeeze bulb fits into the fingertip and connects to the LX0503 through a small piece of tubing. But before you rush out to the local aquarium supply store for air hose, read on.

Hygenic Corporation (1245 Home Avenue, Akron, OH 44310) manufactures latex sheeting, plastic, rubber, and latex tubing, primarily for medical and industrial applications. They have an amber latex tube with a 3/16-inch inside diameter, 1/16-inch wall, and 5/16-inch outside diameter. Not only does it fit the LX0503 perfectly, but the thin walls distort when squeezed, providing a bellows effect.

To accomplish the bellows effect, of course, both ends of the tube (or at least one) have to be stoppered. The stopper at the fingertip— have you guessed?—is the squeeze bulb from the eyedropper or medicine dropper. A dab of epoxy or cyanoacrylate cement should bond the tubing to the bulb quite readily.

The tubing is then routed along the inside surface of the finger or thumb, just outside the flexor/extensor wire keepers. The tubing ends at the circuit board inside the android's palm-box on which the LX0503 transducers are mounted.

The Omron microminiature switches mentioned at the end of the chapter on collision and obstacle avoidance can provide an additional level of grasp sensing. These are fitted inside the links of chain (they're that small!) and cemented in place—be careful not to cement the actuator button. The button is situated so as to face the tubing, but, of course, the flexor/extensor wire is between them. A small U-shaped piece of any rigid material—plastic or metal, for example—is looped over the wire and its open ends cemented to the outside of the tubing. The bottom of the U then presses in the switch button any time an object in the android's grasp displaces the tubing, independently of the bellows effect of the tube.

Normal flexing of the fingers and thumbs will, of course, cause some pressure reading in the tubing, and possibly some switch closures. The android can be programmed to teach itself what these values are for each reading from the servo position pot on the motor/pulley. This not only gives the android a sense of how hard it is grasping an object, it can alert the android that something's wrong. "My hand hurts," thinks the android. Intriguing, isn't it?

There still should be enough room left in the android's fingertip to squeeze in one more sense—temperature. The sensors are the size of small plastic signal transistors, but they're actually 3-lead adjustable current source ICs. The LM334 (again from National Semiconductor) can provide a voltage proportional to Kelvin temperature with just a positive supply connection on pin 2 and a 220-ohm resistor from pin 1 to pin 3 at the fingertip. A lead from pin 3 is brought back through a signal line to a 10K resistor to ground; the voltage across the 10K resistor, which is located near the A/D converter that feeds the processor interested in temperature data, is proportional to temperature and approximately 10 millivolts per degree Kelvin—between roughly 2.7 and 3.0 volts for most temperatures of interest.

One other approach to fingertip senses we should mention, although one that I have no idea how or why to process, is a *tactile* sense. Les Solomon of *Popular Electronics* suggests this demonstration of his unique approach.

Take an inexpensive phonograph cartridge, a small amplifier, and a small speaker—3 inches, more or less. Have a friend drag the stylus over a number of surfaces while you're blindfolded. With your finger on the cone of the speaker, you'll find you can identify any number of textures. A small enough cartridge in the fingertips can do the same for the android, I suppose. Try it if you're interested.

That's about it for the senses I know how to pack into the android's fingers, the bones, the muscles, the works. Now all the android needs is skin.

You may have read of a Russian development where carbon granules were packed between electrodes that were sandwiched between layers of rubber. Like the carbon microphone in a telephone, when compressed, this skin exhibited a variable resistance that was used to provide a voltage that varied with pressure. No doubt this worked under laboratory conditions. But carbon granules exhibit a packing phenomenon, which makes this approach questionable for actual android applications.

I understand that a variation on this approach in the United States substituted conductive foam of the type used to static-protect integrated circuits during shipment and handling. Having had no experience with this approach, I can only suggest it for your experimentation.

If we try not to accomplish anything electronic with the skin, what properties should it have? Flexibility is a requirement if our pressure sensors are going to be able to do their job. Waterproofing is necessary because the android may have to manipulate things under water (you did want it to wash your dishes, didn't you?) or even just walk in the rain. The surface also should be slightly frictional to allow the android the same kind of gripping ability our own hands have. There's an easy answer to all of these: rubber gloves.

The rubber gloves you buy in the supermarket should do, even though they probably won't fit. Buy several pair. Cut the fingers off and start by fitting them over the ends of the android's fingers and thumbs. Do the same again, this time cutting the ends off the fingers and using the tubes that remain, slide these down to the bottom of the fingers and thumbs, allowing about an inch of overlap. Use a bead of cement or silicon caulk to seal the joint. Cover the remainder of the hand with either the large remaining pieces of glove, sealing as you go, or buy latex sheeting for the purpose.

Oops, forgot to mention something earlier when discussing motors. You don't need to include five motors. You can drive the two thumbs together with one motor, the two outer fingers together with a second, and the center finger with a third and still provide the android with a grip capable of accomplishing just about anything short of playing a muscial instrument.

The *manipulative imperative* tells us that an android must be able to provide something like hands and fingers in order to qualify as an android. To coexist with us in our environment, it has to be able to maneuver objects designed to be maneuvered by humanlike appendages. But even if we didn't *have* to give the android this capability, you have to admit—it's handy.

16
Vision by Ramera

We are about to embark on what is certain to be the most talked about chapter in this book. Because you are about to learn about a 64 × 64 pixel digital camera the size of your fist that you can build for about $20.

I call this amazing imager a "ramera" because the active imaging element is a dynamic RAM memory chip. (Yes, that is a redundant expression since RAM stands for "random access memory"—so we've called it a dynamic random access memory memory chip. This confession has been sponsored by the *Electronic Engineering Times* Picky People Patrol, chaired by Instrumentation Editor Chuck Small; we now return you to the overly dramatized relevation, in progress.)

The ramera was first developed by the Robotics group at Case-Western Reserve University in Cleveland, Ohio, in 1978, which included Dave Fotland, Steve Chalmers, Seth Hefter (who formed the Motion Platform Group); Gary Coleman, Mark Fridell, Robert Frederking, Mark Neman, and Gregg Ordy (who formed the Sensor Group). Here is a passage from the Sensor Group journal, kindly forwarded by Gary Coleman:

A Surprise Development—Coleman and Ordy discuss an old article about a dynamic RAM TV camera. Although not specified in the article . . . the RAM is a 4008 with the lid popped off. Gregg gets an article about the *Cyclops*, and will build it if Coleman can get the lid off of his RAM. Coleman takes a file to it and finds that once one edge is gone, a swift blow will knock the lid off. Much to everyone's surprise and delight, it appears that it is quite easy to avoid breaking wires, and that the air is not corrosive in the short run. Ordy builds a circuit that displays the picture on a scope, and this convinces everyone that this is a worthwhile thing to put on our robot. It would require its own processor to accomplish anything useful (it would be a real challenge even with a dedicated PDP-10).

Neman presents a lecture on scene analysis. Apparently there are two ways to find objects—find a point and keep enlarging

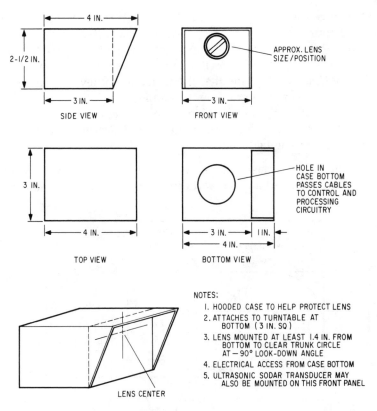

Fig. 16-1 *Camera box (two places).*

it until you hit an edge or locate all the edges in the picture. It's not clear how you pick between them.

The camera now has a lens, and people viewing the oscilloscope image with only one bit of gray scale can recognize individual faces (close up). Seth Hefter now gives a lecture about visual perception. A 16 by 16 array with enough levels of gray scale is adequate to allow human recognition of individuals' faces. We have a 32 by 32 array, with an undetermined number of gray levels, so we should be ok.

The "Cyclops" article referred to was entitled "Build CY-CLOPS—First All Solid-State TV Camera for Experimenters" when it first appeared in *Popular Electronics,* and more recently in the 1978 edition of their *Electronic Experimenter's Handbook.* Authors Terry Walker, Harry Garland, and Roger Melen describe the imager only as a "MOS array" in a 16-pin DIP with a transparent window, available from a private source for $50. Cyclops counts through the 1024-bit array with four 7493 four-bit binary counters, which also drive discrete resistor ladder networks in a crude D/A converter scheme to provide horizontal and vertical voltages to an oscilloscope, presenting a counted raster scan. Gary Coleman tells me that the raster was uneven as all get-out, but the images were recognizable (to humans).

Gary offers this description of the ramera in an article, "Eyes for Your Computer," which appeared in *Shift Register*, the newsletter of the Cleveland Digital Group:

> The Robotics class at Case, of which Gary Coleman is a member, has developed a quick and dirty image sensor. It's almost free.

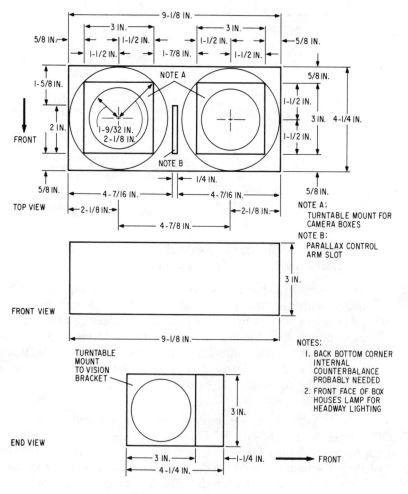

Fig. 16-2 *Camera mount.*

Take a 1K dynamic RAM chip—the MK4008 family is best because it's pin compatible with its static brother, 2102. Carefully remove the gold lid on the chip so that the memory array is exposed. Be sure that you haven't inadvertently broken or shorted any of the tiny wires that connect it to the outside world. If you focus an image on the memory array, write ones into all locations in the chip and then read it, you'll find that where the light was bright, there is a zero; where the image was dark, the ones will still be there.

This works because of the photoelectric effect, which says that if light shines on an object, it will knock electrons off of it.

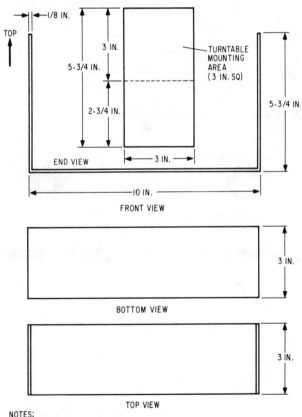

NOTES:

1. MOUNTS AS FAR FORWARD AS POSSIBLE ON NECK DISK (19 IN.); INTERCEPTS ARC OF DISK (CIRCUMFERENCE) AT FORWARD EDGES

2. TURNTABLES (3 IN. x 3 IN.) MOUNT INSIDE UPRIGHTS, HIGHEST POSSIBLE POSITION, THREE EDGES FLUSH

3. HIGH MOMENT-ARMS REQUIRE RIGID MOUNTING OF BRACKET BASE TO NECK DISK

Fig. 16-3 *Vision bracket.*

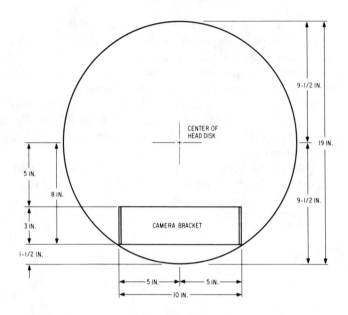

Fig. 16-4 *Top view of head disk showing position of camera bracket.*

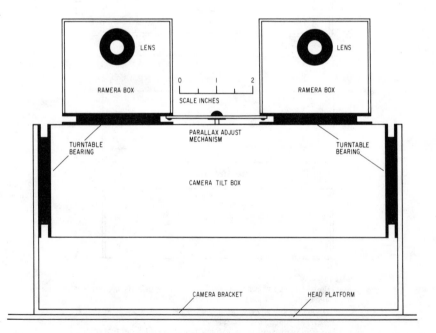

Fig. 16-5 *Front view of camera assembly showing rameras, camera tilt box, and mounting bracket while looking straight forward (0° tilt).*

The rate at which the electrons are released is related to the intensity of the light. This principle makes it possible to achieve gray levels, if you wish, by simply sampling the memory a few times. The first time you sample it, only the brightest parts of the image will be visible. Scan again an instant later and the next brightest part will now be visible.

The 4008 is logically compatible with the 2102 with just two exceptions: first, it's dynamic (which is essential for this to work!!), and second, it uses −12 Volts on the ground pin. You will probably want to build a sense amplifier that will allow you to set the contrast (one transistor). Just hang this chip on your memory bus and maybe you can develop some software that would let your computer read text out of a magazine or a typewritten page.

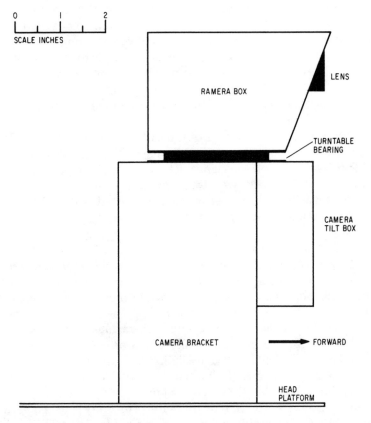

Fig. 16-6 *Side view of head showing position of rameras, camera tilt box, and mounting bracket while looking straight forward (0° tilt).*

Gary also forwarded some notes on the prototype ramera that are interesting. He suggests that you do not use ripple counters, since they ripple through meta-states that make the beam on an oscilloscope hop around in a few places; instead, use synchronous counters. Some memory chips are not wired up to scan addresses in order in a simple raster progression. Experiment with swapping, inverting, or logically interpreting address lines. And "don't fool around with resistors. Go buy an 8-bit D/A—they're cheap!"

That's the seed of experimentation that will hatch affordable vision for our android and countless other applications. But let's take a look at some of the refinements we can add to our ramera.

First, memories have become less expensive, and it's now quite reasonable to buy a 4K dynamic RAM for 64×64 bits of picture resolution. The 16-pin 2104 is okay, but the 22-pin 2107 is probably easier to use since all address lines are brought out to pins individually instead of multiplexed.

Secondly, even for one bit of information, there are two ways of determining the threshold at which a cell would be depleted from a 1 to a 0. One way is to vary the delay between writing 1's into each location, then reading the memory. The other way is to bias the memory with light. Here, a small pilot lamp provides some illumination of the face of the RAM but not quite enough to swing the cells low; only a small amount of incident light from the focused image is then necessary to make the cell deplete.

One aesthetic advantage of this latter scheme is that the darker the ambient light, and therefore the image, the brighter the biasing light. The result is an effect science fiction has been suggesting for generations—an android with eyes that glow in the dark.

The biasing light can easily be constructed with a simple photoresistor or phototransistor determining the voltage or current at the lamp.

What we've seen so far are the basics of the ramera. We still have a way to go before we've provided eyes for the android. There are mechanics to be constructed, software requirements to be examined, and more. So shift these basics to the back burner, and let's take a look at how to turn an element into a system.

The ramera is not your only camera alternative. Periphicon (P.O. Box 324, Beaverton, OR 97005) offers a 32×32 pixel (short for "picture element") Optical Image Digitizer, Type 511, for about $200. Hughes Aircraft Company (Industrial Products Division, 6155 El Camino Real, Carlsbad, CA 92008) has begun offering Omneye® charge-coupled imagers. The HCCI-100A 100×100 pixel array is priced between $550 and $1,400, depending on quantity and grade. The HCCII-032A is a 32×32 pixel array priced between $350 and $500. These devices require

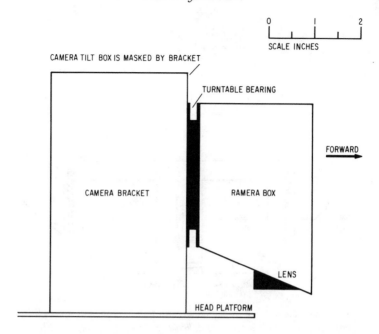

0 1 2
SCALE INCHES

CAMERA TILT BOX IS MASKED BY BRACKET

TURNTABLE BEARING

FORWARD

CAMERA BRACKET

RAMERA BOX

LENS

HEAD PLATFORM

Fig. 16-7 *Side view of head showing position of rameras, camera tilt box, and mounting bracket while looking straight down (−90° tilt).*

a four-phase clock, but include on-chip output amplifiers capable of driving a 20-picofarad load at up to 3 megahertz. And a 64×64 pixel photodiode array is available for $860 from Integrated Photomatrix Inc. (1101 Bristol Road, Mountainside, NJ 07092). None of these are quite as inexpensive as a ramera, but you may prefer not breaking up a few RAMs in order to get the cap popped properly off of two of them. (By the way, I found a careful soak in nail polish remover helps finesse the cap off without a sharp rap—my score so far is two good out of three tried.)

Cyclops was named after the mythological one-eyed giant. This big fellow was something of a curiosity because nature doesn't provide us with many examples of nonbinocular vision. Binocular (two-eye) vision offers us the advantage of depth perception, which develops into an ability to integrate visual information into perceptions of the positions of objects in space (yes, you relativity buffs, in time, too).

The relatively low cost and small size of the ramera lets us allow the android the luxury of binocular vision; now it's up to us to design a visual system that lets the android take full advantage—easily—of the advantages of binocular vision.

We'll start by designing a baseline triangulation system, which is easier than it sounds (see Figs. 16-1 through 16-8). A couple of Ed-

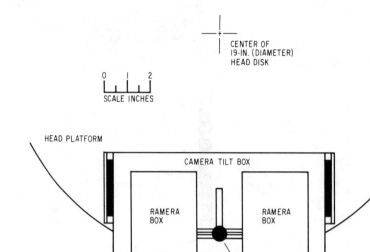

Fig. 16-8 *Top view of head showing position of rameras, camera tilt box, and mounting bracket while looking straight forward (0° tilt).*

mund's 3-inch turntable bearings let the ramera boxes pivot radially. If we tie together the front edges of these turntables (more precisely, the front inside corners), a simple linear sliding action swivels the cameras in tandem. The cameras need to swing through a range of 0 to 90° relative to the baseline. Simple trigonometry applied to this classical parallax example tells us the range to an object colocated at the same point in the visual field of two identical cameras can be calculated as the product of half the baseline length times the tangent of the angle of the camera relative to the baseline (see Fig. 16-9).

Let's take a quick look at some of the hardware this involves. First, the linear actuator that adjusts the camera parallax has become very easy to accomplish, thanks to Airpax (North American Philips Controls Corporation, Cheshire Industrial Park, Cheshire, CT 06410). Their Series 92100 Digital Linear Actuator (Fig. 16-10) and companion Series SAA1027 Stepper Motor IC Driver are also available from Philips' subsidiary Signetics and take 12-volt CMOS-compatible logic level inputs on set, trigger, and direction pins, translating them directly into variable extension of the stepper's shaft. In quantity, suggested resale is $11.00 for the motor and $4.30 for the IC; in units, you might have to pay over $50 for the pair, so check with the local Airpax sales representative for someone you may be able to buy through at something closer to the quantity price.

You can sense the parallax angle directly through a slider pot or linear transducer coupled to the actuator, a standard pot coupled to either turntable, or indirectly through counting steps to the actuator. No matter how this angle is determined, a lookup table in memory can translate the digitized resistance (or whatever) directly into a range figure.

This works best with the rameras in focus, but can work even when they are not. The vertical columns at the center of the picture array in each ramera are compared to each other; when they match, the range to the edge being viewed is read. That's the theory, but in practice, some refinement is necessary.

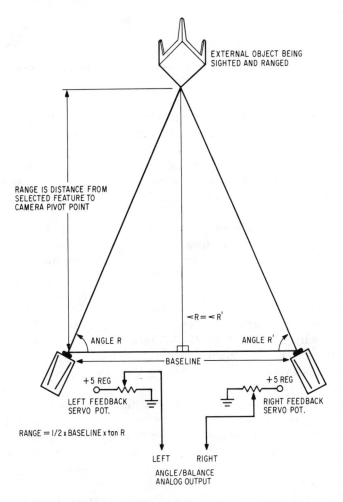

Fig. 16-9 *Application of parallax angle in determining range.*

Fig. 16-10 *Linear Digital Actuator from Airpax. This is a stepping motor with linear rather than rotary motion as its output. (Courtesy of Airpax)*

The reason for this is that in any vertical slice of picture at any instant, the android may be seeing a number of objects at different distances. By definition, the parallax difference in perspective from the two ends of the baseline will place these several objects in different vertical columns in the two rameras. So, unless the object being ranged is tall enough to cover the entire height of the viewed image (or, in a more sophisticated system, the majority of the height), this won't work.

A much better approach is to confine our target to a "crosshair" position at the center of the array. For example, for a 64×64 pixel array in a 4K memory, we can examine eight specific locations—all in the 32nd column, in rows 28–35. These left and right ramera "focus data" bytes can be subtracted from each other, with a 0 result signifying that we've hit a match.

A similar system is used by Honeywell (Visitronics Group, P.O. Box 5227, Denver, CO 80217) in the Visitronic Auto/Focus system, which is appearing on selected cameras. This system uses a custom proprietary IC (sorry, not available without signing a licensing agreement—which costs more than an android—unless you want to buy a camera with the chip in it and throw the camera away) with twin multiple-photodiode arrays and maximum-peak sensing correlation circuitry on chip. Images from one fixed and one rotating mirror, the latter mechanically linked to the lens position in its focusing rack, are reflected by a prism and focused on the chip through twin lens systems in the module. The entire module is about the size of a flashcube.

What about focusing an image on the ramera? Forget it. We're going to take the approach used in cheap box cameras and use a short,

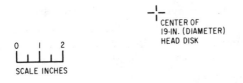

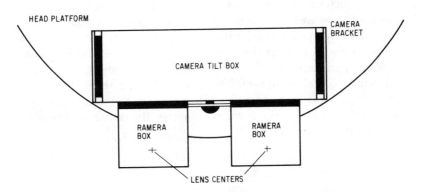

Fig. 16-11 *Top view of head showing position of rameras, camera tilt box, and mounting bracket while looking straight down (−90° tilt).*

wide-angle lens. Virtually everything beyond an inch or so away is at "infinity" for these lenses, so we only have to worry about the initial focusing of the image on the RAM. Be sure to provide some mechanism for rotating the RAM mounting, by the way, since the chip itself is seldom square to its DIP carrier. The important consideration is that the chips be parallel to each other, not necessarily to any external reference.

A good choice for the ramera is 8-millimeter movie camera lenses. I bought a couple from Edmund Scientific for under $10 apiece.

When you build the boxes that house your ramera, it's a good idea to build in a hood to act as an eyebrow. This helps keep rain out of the works and helps prevent accidents that may bang the lens out of position.

We talked about using a twin pivot mount to accomplish parallax aiming. This twin pivot mount should itself be atop a box that lets the cameras tilt up or down. This, in turn, should be mounted to the rotating head platform. Properly designed, this will let the android look straight down his own trunk, look up, turn to any angle, and have complete visual freedom (see Fig. 16-11).

This also lets us locate objects in both relative and absolute space. (Relative space is calculated in relation to the android, then translated into the android's map of its environment as absolute space.)

It all begins with a range calculation, made by the combined contributions of the visual system parallax facility and the SODAR mounted on the camera tilt box. Note that this permits the android a capability for ranging objects seen through solid transparent surfaces, like windows and glass doors.

The range calculation must be translated into distance from the android, which can be easily accomplished by reading both the ramera height above ground (a fixed figure less a lookup table figure addressed by the heads-up system trunk tilting motor servo lean-angle and plumb-angle pots) and the tilt of the camera. This also gives the height above ground for each horizontal row of image pixels at any given range and lets the android calculate both height and altitude at less than the "infinity" parallax position. (An absolute value for this limiting range depends on the length of the parallax baseline, the distance between centers of the imaging elements.)

Fig. 16-12 *High contrast image of the author, processed from small photo by artist John Scavnicky.*

The position of the head platform relative to the android body and the position of the android within the room (or other mapped area) complete the translation to map references. There are two ways of easing these calculations. Either individual lookup table utilities can be provided for each calculation (since these work directly from digitized data already in the system, they are as much as several million times faster than the alternative, though they require a considerable investment in development time and a moderate investment in nonvolatile memory) or a microprocessor-compatible calculator can be incorporated as a math utility.

The National Semiconductor Corporation (2900 Semiconductor Drive, Santa Clara, CA 95051) MM57109 Number Cruncher Interface to Microprocessor IC is a custom version of the company's COP series of microprocessors designed to perform 12-digit scientific calculations, including trig, log, and more, while in communication with a host microprocessor. Unfortunately, instruction execution times range from 1 mil-

Fig. 16-13 *A 32 × 32 "digitized" version of image in Fig. 16-12. This image shown is the basis for Figs. 16-14 through 16-19.*

lisecond to 1 second. It is by far the easier approach for us, but cripplingly slow for the android. Perhaps your own ingenuity can help strike a compromise. The 12-digit accuracy of the MM57109, of course, is hardly necessary for our application. Still, it *is* temptingly easy to implement.

It might make sense to develop a simulator using the digitized data available from the android and a module incorporating the MM57109. A heuristic program could be used to develop the given iterations required for the lookup tables without requiring quite so much of your own time.

So far, we've only made minimal use of the ramera's visual capabilities. These little beauties do, after all, give us images. We might well take the raw images themselves and compare them to data templates in an effort to recognize a limited number of significant objects, but a few simple manipulations of the data may help.

The image as it "appears" at the ramera is extremely high in contrast. While we could accomplish any given number of gray scale iterations with time, this high-contrast one-bit image will prove much easier to handle—and, in the sense of processing, it's free.

A series of logical iterations can find the edges of significant solid areas. Figures 16-12 through 16-19 help explain how this is accomplished.

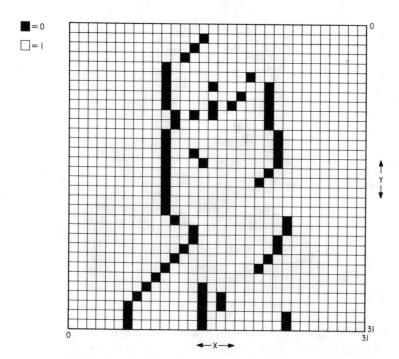

Fig. 16-14 *Left side outline results from logical operation: LET D(X, Y)=0 IF D(X,Y)=0 AND D (X −1,Y)=1, ELSE LET D(X,Y)=1.*

□ = 1
■ = 0

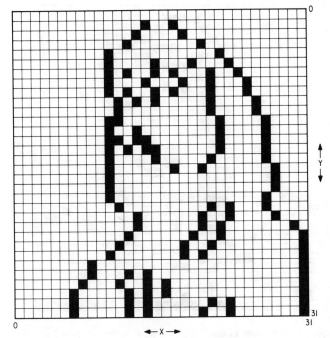

Fig. 16-15 *Right side outline can be added to left by modifying the logic to: LET D(X,Y)=0 IF D(X,Y) =0 AND (D(X −1,Y)=1 OR D(X+1,Y)=1), ELSE LET D(X,Y)=1.*

■ = 0
□ = 1

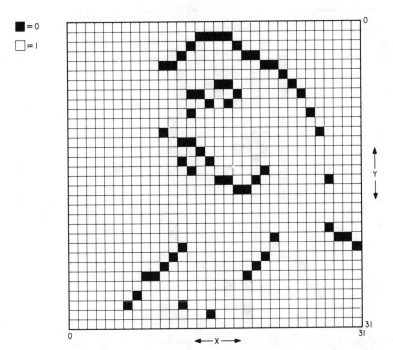

Fig. 16-16 *Top side outline results from logical operation: LET D(X,Y)=0 IF D(X,Y)=0 AND D(X,Y −1)=1, ELSE LET D(X,Y)=1.*

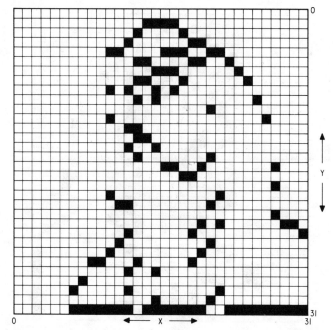

Fig. 16-17 *Bottom outline can be added to top by modifying the logical statement to: LET D(X,Y)=0 IF D(X,Y)=0 AND (D(X,Y−1)=1 OR D(X,Y+1)=1), ELSE LET D(X,Y)= 1.*

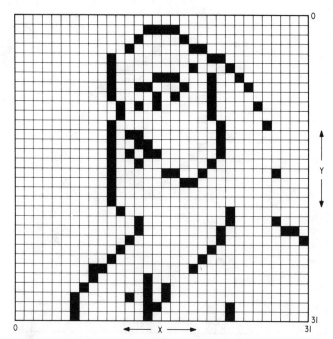

Fig. 16-18 *"Drop shadow" combines all top and left outlines, yields good recognition and tremendous data reduction: LET D(X,Y)=0 IF D(X,Y)=0 AND (D(X−1,Y)=1 OR D(X,Y−1)=1), ELSE LET D(X,Y)=1.*

Scanning from left to right, we can identify all left edges by determining if the bit being examined is a 0 and the bit before it is a 1. Remember, an image is identified by the depletion of a specific cell location from a 1 to a 0.

Scanning from top to bottom, we can identify all top edges by determining if the bit being examined is a 0 and the bit above it is a 1. Following this so far?

To get both left and right edges, we need to know if the bit being examined is a 0 and either the bit before it or the bit after it is a 1.

Similarly, we can identify both top and bottom edges by determining that the bit being examined is a 0 and either the bit above it or the bit below it is a 1.

We could get a drop-shadow effect by identifying that a bit being examined is a 0 and either the bit to its left or the bit above it is a 1. Ready for the big time yet?

We can get a full outline from the initial image by finding every bit that is a 0 where the bit above it or the bit below it or the bit to its left or the bit to its right is a 1.

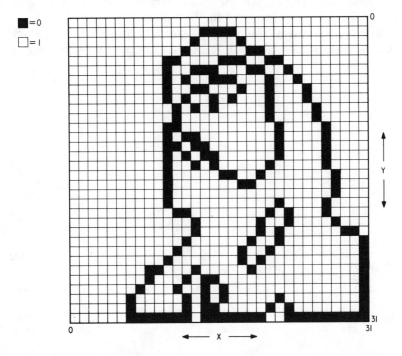

Fig. 16-19 *Full outline results when left, right, top and bottom outlines are combined: LET D(X,Y)=0 IF D(X,Y)=0 AND (D(X −1,Y)=1 OR D(X+1,Y)=1 OR D(X,Y −1)=1 OR D(X,Y+1)=1), ELSE LET D(X,Y)=1.*

If we identify the horizontal address of a specific bit as X and the vertical address as Y, an outline can be produced by comparing the data at surrounding addresses with a logical statement: IFF (X −1=1 AND X=0) OR (X+1=1 AND X=0) OR (Y −1=1 AND Y=0) OR (Y+1=1 AND Y=0) THEN X,Y = 0 ELSE X,Y = 1.

Actually, this is a fudged statement, mixing both data and address terms and leaving out nonessential address identification. For you purists, here's a more formalized statement: LET D(X,Y)=0 IFF D(X,Y)=0 AND (D(X −1,Y)=1 OR D(X+1,Y)=1 OR D(X,Y −1)=1 OR D(X,Y+1)=1) ELSE LET D(X,Y)=1.

The latter part (ELSE, etc.) of the above statement is unnecessary with the IFF (if and only if) statement, but is included here for the sake of clarity.

What we have discovered is an inexpensive way of obtaining an image and reducing it to its outlines. I'm afraid that the art or science (depending on how familiar you are with the task—and you may want to include "magic" as an option) of image analysis is much too ambitious a subject for us to undertake here. That's something you will have to research and experiment with yourself. You'll be in good company. This is a line of research being aggressively pursued throughout science and industry. And they don't have the head start you do—a cheap and simple way of getting the image in the first place.

So let's pour a drink and propose a toast to those helpful fellows at the Case-Western Reserve 1978 Robotics Group: Here's looking at you!

17
Mapping, the Atlas, and Probability Shells

A friend of mine who learned early on in the project that I was preparing a book on android design—a fellow with some knowledge of the subject, but a terrible punster—begged me not to call the chapter on mapping "Grid Expectations."

Androids aren't quite as smart as people. People can integrate their senses into pictures—mental images of where things are, what they are, and so on. But people have a long time to learn these things. If there's an infant in or near your life, you can watch this learning process in progress. Two decades later, that infant will be a fully integrated, fully functional adult.

One of the problems of android design is that our android is hatched full-grown, without the advantage of a long infancy. The best we can do is give the android some quicker means of learning its way around.

This utility is called a *map*, but don't think of it in terms of the Rand-McNally stock in trade, not quite. The map we provide our android with is much more like a blank form it fills in as it goes. The form is a grid-like matrix of locations. In each location, we can store data as to what the android knows is in that location. In its simplest form, that data could be a single bit to indicate whether or not there's an obstacle there.

Let's take a closer look at a simplified map. Let's assume a room perfectly square, 16 feet each way. We could take a 16-square by 16-square section of quadrille graph paper and use it to represent the room, with each of the 256 squares representing a 1-square-foot area within the room. We could use four bits to identify its east-west column, possibly as the lower four bits in an eight-bit address word; the upper four bits to identify its north-south row. The east/west/north/south references here are arbitrary ones. If we assign just one bit of data (1 = something there, 0 = all clear) to each address, we can map four rooms in 1K of memory.

The android could enter the room and methodically explore each of the 256 locations, updating the map data as it finds objects in places where none were recorded or spaces where there previously were objects.

With a second bit available in each data location, the android could increment or decrement the value of the data in each location, converting the score to a level of confidence. A 0 might mean that nothing has ever been observed in a specific location, or that for every time it has been observed there, it also has been observed at least one time not there (which, in this context, means absent and not elsewhere). A 1 might mean that the space has only been observed occupied once, or that it has been observed occupied one time more than it has been observed unoccupied. Similarly, as the value of this score increases through any number of bits, the number of levels of confidence increase.

One mechanism worth considering is to never let the level of confidence decrement to less than 1, and to assign a value of 2 to the increment if the data at the address previously is 0. This removes some confusion from the interpretation of the data at each location. Now, 0 can only mean that nothing has ever been observed at that particular map location. A 1 can only mean that there was something there once, but it isn't there now. A 2 can mean that there is either something there now or a good likelihood of something being there now. And, in a two-bit code, a 3 can only mean that there was something there at least twice, including the last time the location was investigated.

How about that! Just one extra bit brings us so much more information! But is it enough? What others kinds of information do we need? For that matter, how can we use the information we have?

One thing we need to have our androids do is get from one place to another without running roughshod over things in between, to find an efficient path in so doing, and to realize where the place is. Another is to know where to find power outlets so it can recharge when the battery needs it. Another is to know where doors are that let it leave a room.

As we get more and more sophisticated, there are more and more special cases to consider. Perhaps it's time to unsimplify our model.

How big a matrix should a map represent? Assuming one map per room (we could include closets, alcoves, and other small areas as part of rooms near them), how big can a room be? And how big should we make any square in the grid?

In rough terms, the android is about 3 feet long and 2 feet wide. If we were to make the grids 2 feet on a side, then a clear grid square would certainly be enterable, if not traversable, by the android. But a clear area 3 feet wide that happened to straddle two grid sections wouldn't show at all, since the obstacles in the last 6 inches on either side would prompt an "occupied" code for both squares. Reducing the grid square size to a foot each way at least shows a two-square-wide path—and that should be cue enough to the android to get close to the path and use his sensors to check it out, to see if it's wide enough to let him make it through safely.

How many 1-square-foot squares are in a big room? Dealing in the android's world of power of two, is it 8, 16, 32, or 64 feet square? Since we're designing a format for any map we may ever need, 32 seems to be the smallest figure we can use to cover the majority of areas; bigger rooms can always be subdivided and incorporated into multiple maps.

So a ten-bit address can identify any one of the 1,024 byte locations on the map, and a 1K × N memory can store each map, where N is the bit size of the byte. We'll take a look at the data byte at each address in a moment, but first, let's unrandomize the orientation of the room in memory by setting up a few rules.

First, the SODAR-ranging collision-avoidance sensors can easily identify the square dimensions of the room from virtually any position in the room. The android rotates its trunk slowly, taking ranging readings as it goes, but without changing its position in the room. When the readings have all reached a minimum value (assuming a clear path to the wall from each transducer at that position, which is easily ascertainable during the rotational scan), the transducers should be square to the four walls. This can be confirmed by continuing the scan through an additional 90° angle. The sum of the ranges of opposite transducers (plus the diameter of the trunk) gives the dimensions of the room. Since the SODAR module provides a period output and measurements of distance are made by counting oscillator cycles, units of distance are either arbitrary or can be determined at our (or the android's) convenience. Choosing units either 3 or 6 inches long gives us either six or seven bits at maximum for a 32-foot room—which becomes either seven or eight bits if we allow an "overflow" bit.

The extra bit or two of accuracy lets us round off safely—and in a mapping application, we're best off rounding high.

Once we know the dimensions of the room, we can reduce the number on possible permutations of *outlook* to two simply by mandating that our view of the room will always depict it as wider than it is long (or deep); by permutations, we mean that the same room may be presented in a variety of ways in a map simply by varying which end is *up*. To keep our view of the room consistent, we will always start the room with one corner at the upper left corner of the map "page," which we will define as vertical column 0, horizontal row 0.

How do we resolve the *room* versus *upside-down room* map uncertainty? There are a couple of ways to approach this problem. One involves a map-to-map index at each door, the other a sequential search for feature matching in same-size rooms listed in a cross-reference table. Let's take a closer look at these schemes, one at a time.

There are only a few ways a robot can enter or exit a room. Stairways, doorways, and lifts are obvious. Less obvious is the "unconscious entrance," which happens any time the robot "awakens" in a room with no data in its map memory—whether it's because lightning has just hit the rod atop Frankenstein's castle breathing life into the beastie or because it has suffered a hardware failure.

In the event of an unconscious entrance, it isn't crucial that the android resolve the parity. Any given instruction can assign one of the two diagonally opposite wide-over-long corners to 0,0; the corner closest to straight ahead or left, for example, will always give the android *something* to do when faced with that first map situation, and let it begin mapping.

Any time a first map is made, a number of reference table entries are made. First a by-size file 32 bytes long records how many rooms have been mapped for each of 32 values of width, and the appropriate byte here would be incremented by 1.

Second, an entry will be made in a five-byte-per-entry index file. The first byte records the width of the room, the second byte the length, the third and fourth the starting address of its map in memory, the fifth the number of doors (and other entrances and exits) in the room (lower four bits) and the number of power outlets (upper four bits).

Third, a block of 32 bytes is reserved in a door-index block of memory, 2 bytes each for 16 reserved door listings. The first byte gives the location of the door on the map of the room, dropping the least significant bit from the row and column values and storing them in the lower and upper four bits of the byte, respectively. The second byte gives the index file starting address for the room on the other side of the door, if known; if unknown, this is a 0.

These files are updated and expanded each time a new room is added to the android's experience.

The first time an android enters a new room, the entry can be assigned to either the top or left wall (depending whether it is the long wall or the wide wall, as we discussed earlier). Any subsequent time the android enters that room, the door-to-door room-to-room lookup tables "name" that room for the android, and right-side-up is identified by the door it entered the *first* time.

Let's assume that the android isn't foolproof and for some reason or another loses its place while cross-referencing between rooms. Here it is in a new room. Has it been here before?

Okay, that's a reasonable question. What is there about the room it might be able to recognize? Size is a pretty good first check. The android performs the SODAR-sizing operation we described a bit ago and comes up with length and width values for the room. Has it been in any rooms this wide before? The by-size file tells it how many. Now the android checks through the index file. If the width doesn't match, he goes to the next entry. If it does, he checks its length. If this doesn't match, he decrements the number of this-width rooms (fetched from the by-size file) by 1 and moves on; if it does, he decrements the this-width number and stores the map address in a scratch pad section of RAM (perhaps a stack).

This procedure continues until the this-width number has reached 0. The android then fetches the first map listed and checks to see if a given, easily identifiable feature of that map is located in the room he's now in. You can set up a scoring procedure, figuring some minimal N number of matches out of M attempts to be a match. Each map listed in the scratch pad is examined, and in most cases a match will be found. If no match is found, the android must assume itself to be in a new room.

It's up to you to decide how rigorous a procedure this needs to be. In a home, 2 or 3 out of 4 might be a high enough score. In a school with lots of identically proportioned rooms all looking pretty much the same, the check should be made more thoroughly.

Several minutes spent recognizing a room and fetching its map is a very worthwhile investment of the android's time, since remapping a room requires so much more of it.

The total collection of maps accumulated by the android is its "atlas," and we have shown how it can be indexed. Of course, you are encouraged to disregard this scheme entirely and develop your own, to modify it, or to try it.

The last aspect of mapping we want to cover here is one of how data might identify what the android will find at any given map location. You will soon see that by permitting an eight-bit byte at each map location, we can permit the android tremendous flexibility in "knowing" its immediate environment. Also, the relatively small map size (and the same kind of map location LSB-dropping that compresses the ten-bit location into an eight-bit code for the door-index) lets us perform simple serial content (data) searches for addresses where given items might be found.

Here's one suggested scheme for identifying the nature of an object, its height, and the probability of it being where the map says it is.

That latter concept, which we'll call a probability shell, takes into account that things can and often do move around, or are moved around. A table, for example, is highly likely to be in one place and one place only, although it is possible for the table to *not* be there. A light switch, window, door, or outlet is much more likely to stay put and has a very low probability of appearing in any adjacent position. A wastebasket, on the other hand, may be where it was last seen, but there is a significant probability that it will appear in some one of the nearby grid positions, as well as some smaller probability that it won't be anywhere in the room. Pets and humans are very mobile, and there is a high probability that they will not be where they were last observed.

Knowing that we have an eight-bit data word available, let us hereby reserve one bit as a probability-increaser.

Also, while the android's map occupies only two dimensions of array (in the colloquial sense), it traverses three dimensions of space. Height alone is an insufficient parameter, since objects may be at any height or hang down. So let's assign three more bits—one each to correspond to things found at high, middle, and low "altitude" levels, which allows eight basic vertical position (or Z-axis) combinations.

The remaining four bits can be used to categorize, classify, or name the things the android finds, and permits a very comprehensive use of the available data. But what 16 things or kinds of things are significant enough to merit their own categories?

Power outlets are very important, since they mean survival for the android. It's probably a good idea to give them a hex value of F (binary 1111), since this could permit fast bit-ANDing in addition to conventional techniques.

The android must also be aware of where humans are, since it is mandated that safety to humans is a requirement for android behavior and a definite factor in decision making. Similarly, there is a slightly lesser requirement that the android be aware of where animals are and safeguard their well-being.

Property is the third class of things the android must be aware of. Furniture and fixtures are the primary classifications of property with which a map might be concerned.

Of course, the android has to know where walls are located, and where they aren't. Walls ought to be given their own category. Corners also should be given a category, not only because their positions are of interest when identifying a room, but because they can cause unusual reflections of sensory signals, and the possibility of sensory error at specific locations needs to be recognized.

This possibility of sensory error also suggests that mirrors be included as a separate category to keep the android from regarding the room-through-the-looking-glass as real and accessible.

Doors, as we've seen, are a very special category. There are only a few classes of exits other than doors. Of these, stairs, steps, and ramps deserve their own category—especially in light of the power requirements they entail. By the way, you might want to mandate that any room map that includes a stairway should be oriented with the stairway as far away from the 0,0 reference as possible, which should allow it to extend beyond the walls of the room. The other classes of exits (holes in the floor, magic ropes, dimensional portals, etc.) are rare and mostly unusable by the android so should not be put in the same category as doors; still, the android should be able to recognize places where humans and animals can enter and exit, like dumb waiters, windows, and so on.

Since the manipulative imperative means the android has to be able to do things, and since the visual system requires light from standard sources, where possible, to keep the android from running down its battery by keeping its headlight on too long, power and light switches become an important category.

There are, believe it or not, *six* categories of space we want to categorize (which may or may not be empty).

Initially, all space in a map is unclassified or unknown. Specific areas may remain unclassified or unknown in one of two ways. Either there is something there that the android doesn't recognize or the android, for one reason or another, doesn't know what's there.

There is another category that might be confused with this—a classification of spaces and objects that have no special qualification. This includes clear spaces, objects the android recognizes that aren't included in one of the other categories, and objects the android may not recognize but has decided (for whatever reasons you program) not to be curious about.

A third category includes narrow spaces and objects, those that take a maximum of two grid spaces in either the X or Y axis but are not otherwise categorized.

A fourth category includes those spaces and objects the android knows to be dangerous or hazardous, either to itself or to the safety and well-being of humans, animals, or property. It is this category that permits the android to respond to its surroundings in a way that is consistent with the behavioral guidelines we have set down for it, analogous to the "Laws of Robotics." *How* the android determines something belongs in this category is another matter—one which will be a crucial area of further design, experimentation, and development; for the time being, assume that you're going to have to find a way to tell the android when something belongs in this category.

Thinking about this, there may be places and things that are not necessarily dangerous, but where you'd rather not have the android be or meddle if it's at all avoidable. An "avoid this" category will keep the android away from these in most cases but leaves it the option of overriding your request if there is some compelling reason for doing so.

Again, this determination is one that presents an important area for further development.

The last category takes away the android's options and strictly forbids it from entering the space or manipulating the object.

Tables 17-1 and 17-2 present one recommended format for categorizing these several qualifications into an eight-bit data code. As before, you are welcome to hate it and develop your own.

You can see that the mapping requirement is fairly memory-intensive; you also can see that the investment can be very worthwhile. But let me tickle a couple of more ideas into your mind.

Table 17-1 *Map Data.*

$D_8D_7D_6D_5$ $D_4D_3D_2D_1$

D_1 Usually corresponds to low obstacle
D_2 Usually corresponds to middle-level obstacle
D_3 Usually corresponds to high obstacle
D_4 Usually used as a probability increaser

$D_8D_7D_6D_5$ Used as naming/classifying code with these categories:

Binary	Hex	Category
0000	0	No special qualification
0001	1	Narrow (max two addresses in x or two addresses in y)
0010	2	Furniture and fixtures
0011	3	Animal
0100	4	Human
0101	5	Unclassified or unknown
0110	6	Dangerous or hazardous
0111	7	Avoid this
1000	8	Forbidden
1001	9	Stairs, steps, and ramps
1010	A	Mirror
1011	B	Door
1100	C	Wall
1101	D	Corners and exits
1110	E	Switches
1111	F	Outlets

Table 17-2 *Map Data Codes.*

Binary	Hex	Category
0000 0000	00	Clear, free path
0000 0001	01	Possible low obstacle
0000 0010	02	Possible mid obstacle
0000 0011	03	Possible low + mid obstacle
0000 0100	04	Possible high obstacle
0000 0101	05	Possible low + high obstacle
0000 0110	06	Possible mid + high obstacle
0000 0111	07	Possible low + mid + high obstacle
0000 1000	08	Clear but not always clear
0000 1001	09	Probable low obstacle
0000 1010	0A	Probable mid obstacle
0000 1011	0B	Probable low + mid obstacle
0000 1100	0C	Probable high obstacle
0000 1101	0D	Probable low + high obstacle
0000 1110	0E	Probable mid + high obstacle
0000 1111	0F	Probable low + mid + high obstacle
0001 0000	10	Very narrow clear path—go slowly
0001 0001	11	Possible narrow low obstacle
0001 0010	12	Possible narrow mid obstacle
0001 0011	13	Possible narrow low + mid obstacle
0001 0100	14	Possible narrow high obstacle
0001 0101	15	Possible narrow low + high obstacle
0001 0110	16	Possible narrow mid + high obstacle
0001 0111	17	Possible narrow low + mid + high obstacle
0001 1000	18	Sometimes narrow clear path
0001 1001	19	Probable narrow low obstacle
0001 1010	1A	Probable narrow mid obstacle
0001 1011	1B	Probable narrow low + mid obstacle
0001 1100	1C	Probable narrow high obstacle
0001 1101	1D	Probable narrow low + high obstacle
0001 1110	1E	Probable narrow mid + high obstacle
0001 1111	1F	Probable narrow low + mid + high obstacle
0010 0000	20	Space next to furniture/fixture
0010 0001	21	Possible low furniture/fixture
0010 0010	22	Possible mid furniture/fixture
0010 0011	23	Possible low + mid furniture/fixture
0010 0100	24	Possible high furniture/fixture
0010 0101	25	Possible low + high furniture/fixture
0010 0110	26	Possible mid + high furniture/fixture
0010 0111	27	Possible low + mid + high furniture/fixture

Table 17-2 *Continued.*

Binary	Hex	Category
0010 1000	28	Space sometimes next to furniture/fixture
0010 1001	29	Probable low furniture/fixture
0010 1010	2A	Probable mid furniture/fixture
0010 1011	2B	Probable low + mid furniture/fixture
0010 1100	2C	Probable high furniture/fixture
0010 1101	2D	Probable low + high furniture/fixture
0010 1110	2E	Probable mid + high furniture/fixture
0010 1111	2F	Probable low + mid + high furniture/fixture
0011 0000	30	Space next to animal
0011 0001	31	Possible low animal
0011 0010	32	Possible mid animal
0011 0011	33	Possible low + mid animal
0011 0100	34	Possible high animal
0011 0101	35	Possible low + high animal(s)
0011 0110	36	Possible mid + high animal
0011 0111	37	Possible low + mid + high animal
0011 1000	38	Space sometimes next to animal
0011 1001	39	Probable low animal
0011 1010	3A	Probable mid animal
0011 1011	3B	Probable low + mid animal
0011 1100	3C	Probable high animal
0011 1101	3D	Probable low + high animal(s)
0011 1110	3E	Probable mid + high animal
0011 1111	3F	Probable low + mid + high animal
0100 0000	40	Space next to human
0100 0001	41	Possible low human
0100 0010	42	Possible mid human
0100 0011	43	Possible low + mid human
0100 0100	44	Possible high human
0100 0101	45	Possible low + high human(s)
0100 0110	46	Possible mid + high human
0100 0111	47	Possible low + mid + high human
0100 1000	48	Space sometimes next to human
0100 1001	49	Probable low human
0100 1010	4A	Probable mid human
0100 1011	4B	Probable low + mid human
0100 1100	4C	Probable high human
0100 1101	4D	Probable low + high human(s)
0100 1110	4E	Probable mid + high human
0100 1111	4F	Probable low + mid + high human

Table 17-2 *Continued.*

Binary	Hex	Category
0101 0000	50	Unclassified space
0101 0001	51	Possible unclassified low obstacle
0101 0010	52	Possible unclassified mid obstacle
0101 0011	53	Possible unclassified low + mid obstacle
0101 0100	54	Possible unclassified high obstacle
0101 0101	55	Possible unclassified low + high obstacle
0101 0110	56	Possible unclassified mid + high obstacle
0101 0111	57	Possible unclassified low + mid + high obstacle
0101 1000	58	Narrow unclassified space
0101 1001	59	Probable unclassified low obstacle
0101 1010	5A	Probable unclassified mid obstacle
0101 1011	5B	Probable unclassified low + mid obstacle
0101 1100	5C	Probable unclassified high obstacle
0101 1101	5D	Probable unclassified low + high obstacle
0101 1110	5E	Probable unclassified mid + high obstacle
0101 1111	5F	Probable unclassified low + mid + high obstacle
0110 0000	60	Dangerous space
0110 0001	61	Possible dangerous low obstacle
0110 0010	62	Possible dangerous mid obstacle
0110 0011	63	Possible dangerous low + mid obstacle
0110 0100	64	Possible dangerous high obstacle
0110 0101	65	Possible dangerous low + high obstacle
0110 0110	66	Possible dangerous mid + high obstacle
0110 0111	67	Possible dangerous low + mid + high obstacle
0110 1000	68	Sometimes dangerous space
0110 1001	69	Probable dangerous low obstacle
0110 1010	6A	Probable dangerous mid obstacle
0110 1011	6B	Probable dangerous low + mid obstacle
0110 1100	6C	Probable dangerous high obstacle
0110 1101	6D	Probable dangerous low + high obstacle
0110 1110	6E	Probable dangerous mid + high obstacle
0110 1111	6F	Probable dangerous low + mid + high obstacle
0111 0000	70	Avoid this space
0111 0001	71	Avoid possible low obstacle
0111 0010	72	Avoid possible mid obstacle
0111 0011	73	Avoid possible low + mid obstacle
0111 0100	74	Avoid possible high obstacle
0111 0101	75	Avoid possible low + high obstacle
0111 0110	76	Avoid possible mid + high obstacle

Table 17-2 *Continued.*

Binary	Hex	Category
0111 0111	77	Avoid possible low + mid + high obstacle
0111 1000	78	Sometimes avoid this space
0111 1001	79	Avoid probable low obstacle
0111 1010	7A	Avoid probable mid obstacle
0111 1011	7B	Avoid probable low + mid obstacle
0111 1100	7C	Avoid probable high obstacle
0111 1101	7D	Avoid probable low + high obstacle
0111 1110	7E	Avoid probable mid + high obstacle
0111 1111	7F	Avoid probable low + mid + high obstacle
1000 0000	80	Forbidden space
1000 0001	81	Forbidden possible low obstacle
1000 0010	82	Forbidden possible mid obstacle
1000 0011	83	Forbidden possible low + mid obstacle
1000 0100	84	Forbidden possible high obstacle
1000 0101	85	Forbidden possible low + high obstacle
1000 0110	86	Forbidden possible mid + high obstacle
1000 0111	87	Forbidden possible low + mid + high obstacle
1000 1000	88	Sometimes forbidden space
1000 1001	89	Forbidden probable low obstacle
1000 1010	8A	Forbidden probable mid obstacle
1000 1011	8B	Forbidden probable low + mid obstacle
1000 1100	8C	Forbidden probable high obstacle
1000 1101	8D	Forbidden probable low + high obstacle
1000 1110	8E	Forbidden probable mid + high obstacle
1000 1111	8F	Forbidden probable low + mid + high obstacle
1001 0000	90	Space next to stair/ramp
1001 0001	91	Step up
1001 0010	92	Step down
1001 0011	93	Stairs up
1001 0100	94	Stairs down
1001 0101	95	Ramp up
1001 0110	96	Ramp down
1001 0111	97	"Cliff"/impasse
1001 1000	98	Narrow space next to stair/ramp
1001 1001	99	Narrow step up
1001 1010	9A	Narrow step down
1001 1011	9B	Narrow stairs up
1001 1100	9C	Narrow stairs down
1001 1101	9D	Narrow ramp up
1001 1110	9E	Narrow ramp down

Table 17-2 *Continued.*

Binary	Hex	Category
1001 1111	9F	Landing
1010 0000	A0	Space next to possible mirror
1010 0001	A1	Possible low mirror
1010 0010	A2	Possible mid mirror
1010 0011	A3	Possible low + mid mirror
1010 0100	A4	Possible high mirror
1010 0101	A5	Possible low + high mirror
1010 0110	A6	Possible mid + high mirror
1010 0111	A7	Possible low + mid + high mirror
1010 1000	A8	Space next to probable mirror
1010 1001	A9	Probable low mirror
1010 1010	AA	Probable mid mirror
1010 1011	AB	Probable low + mid mirror
1010 1100	AC	Probable high mirror
1010 1101	AD	Probable low + high mirror
1010 1110	AE	Probable mid + high mirror
1010 1111	AF	Probable low + mid + high mirror
1011 0000	B0	Space next to door
1011 0001	B1	Possible low door
1011 0010	B2	Possible mid door
1011 0011	B3	Possible low + mid door
1011 0100	B4	Possible high door
1011 0101	B5	Possible low + high door
1011 0110	B6	Possible mid + high door
1011 0111	B7	Possible low + mid + high door
1011 1000	B8	Space next to narrow door
1011 1001	B9	Probable low door
1011 1010	BA	Probable mid door
1011 1011	BB	Probable low + mid door
1011 1100	BC	Probable high door
1011 1101	BD	Probable low + high door
1011 1110	BE	Probable mid + high door
1011 1111	BF	Probable low + mid + high door
1100 0000	C0	Space next to possible wall
1100 0001	C1	Possible low wall
1100 0010	C2	Possible mid wall
1100 0011	C3	Possible low + mid wall
1100 0100	C4	Possible high wall
1100 0101	C5	Possible low + high wall
1100 0110	C6	Possible mid + high wall

Table 17-2 *Continued.*

Binary	Hex	Category
1100 0111	C7	Possible low + mid + high wall
1100 1000	C8	Space next to possible wall
1100 1001	C9	Probable low wall
1100 1010	CA	Probable mid wall
1100 1011	CB	Probable low + mid wall
1100 1100	CC	Probable high wall
1100 1101	CD	Probable low + high wall
1100 1110	CE	Probable mid + high wall
1100 1111	CF	Probable low + mid + high wall
1101 0000	D0	Space next to possible exit to room/hall
1101 0001	D1	Space next to possible exit to outside
1101 0010	D2	Space next to possible other exit
1101 0011	D3	Possible exit to room/hall
1101 0100	D4	Possible exit to outside
1101 0101	D5	Possible other exit
1101 0110	D6	Possible convex corner
1101 0111	D7	Possible concave corner
1101 1000	D8	Space next to probable exit to room/hall
1101 1001	D9	Space next to probable exit to outside
1101 1010	DA	Space next to probable other exit
1101 1011	DB	Probable exit to room/hall
1101 1100	DC	Probable exit to outside
1101 1101	DD	Probable other exit
1101 1110	DE	Probable convex corner
1101 1111	DF	Probable concave corner
1110 0000	E0	Accessible space next to switch
1110 0001	E1	Possible low switch
1110 0010	E2	Possible mid switch
1110 0011	E3	Possible low + mid switch(es)
1110 0100	E4	Possible high switch
1110 0101	E5	Possible low + high switch(es)
1110 0110	E6	Possible mid + high switch(es)
1110 0111	E7	Possible low + mid + high switch(es)
1110 1000	E8	Largely inaccessible space next to switch
1110 1001	E9	Probable low switch
1110 1010	EA	Probable mid switch
1110 1011	EB	Probable low + mid switch(es)
1110 1100	EC	Probable high switch
1110 1101	ED	Probable low + high switch(es)
1110 1110	EE	Probable mid + high switch(es)
1110 1111	EF	Probable low + mid + high switch(es)
1111 0000	F0	Space next to probably free outlet
1111 0001	F1	Possible low outlet

Table 17-2 *Continued.*

Binary	Hex	Category
1111 0010	F2	Possible mid outlet
1111 0011	F3	Possible low + mid outlet(s)
1111 0100	F4	Possible high outlet
1111 0101	F5	Possible low + high outlet(s)
1111 0110	F6	Possible mid + high outlet(s)
1111 0111	F7	Possible low + mid + high outlet(s)
1111 1000	F8	Space next to probably occupied outlet
1111 1001	F9	Probable low outlet
1111 1010	FA	Probable mid outlet
1111 1011	FB	Probable low + mid outlet(s)
1111 1100	FC	Probable high outlet
1111 1101	FD	Probable low + high outlet(s)
1111 1110	FE	Probable mid + high outlet(s)
1111 1111	FF	Probable low + mid + high outlet(s)

Consider, if you will, that the atlas maintained in memory is a reference utility, and the processor that maintains this atlas and its corresponding memory need not do anything else. A separate processor can be responsible for assigning data categories and will need a great deal more program memory than any other kind. There also is a tremendous opportunity here to incorporate some heuristics into the program to let the android learn how to refine its decisions as it gains experience. This processor will need to communicate with the visual and sensory processors.

The visual and sensory processors also will need to communicate with a processor capable of translating relative positions into map addresses.

In fact, it's probably a good idea to fetch a map from the atlas and store it in a reserved memory space in a current-situation processor. A second memory space in this processor might translate the fetched map into a self-centered map (one where the android is always at the central coordinates) which is always rotated so that the android is always facing in the same direction as presented in memory and the grid locations of the map are all that change—in other words, a heads-up display. This utility will make it much easier for the android to translate between its sensory and visual systems to its mapping system.

Processor organization is a topic we'll be covering in another chapter.

This chapter, at least, has given us a look at how to give the android a collection of maps it won't have trouble learning how to fold.

18
Voicebox

These days, everything talks—games, translators, calculators, wristwatches, television sets, and so on. Speech synthesis is everywhere and available for our android.

Even your telephone, strictly speaking (no pun intended), has become a speech synthesizer. Because electronics breaks down voices into pulse patterns, multiplexes them into complex channelized transmissions, demultiplexes them, restores them into voices, and delivers them to your ear. That's about the neatest trick since freeze-dried coffee—except the telephone trick, historically, came first.

The speech synthesizers we think of when we use the words are almost invariably associated with computers. The only real difference between these and the phone company's is that instead of transmitting the coded pulses, they're stored in memory.

The essential difference between speech synthesis schemes is in how these pulses are coded from speech (and decoded back into speech later).

The object of these various coding schemes is to store the speech in the minimum possible amount of memory and to use the minimum possible amount of memory in decoding it—or other hardware, for that matter.

Another goal that we may want to set for ourselves is some minimum vocabulary requirement. By the way, no matter what scheme we build or buy, we'll almost invariably have to add circuitry to it in order to get it to say what we want when we want. Deciding when something should be said—and what—will require yet another level of intelligence.

For right now, we'll confine our discussion to the speech synthesizer itself, the *voicebox* of the android.

Of the many coding schemes available for speech synthesis, the one with the most promise for virtually unlimited vocabulary with reasonable hardware requirements reproduces the phonemes of human speech. These phonemes are the elementary sounds we utter—like ch, sh, th, nn, mm, oh, ee, ih, k, etc.—and combine into words. A number of *phonemic* (or *phonetic*) synthesizers are available commercially, and we'll take a look at a few of them in a minute.

Unfortunately, the art of phonemic synthesis is still imperfect. Many of these synthesizers sound mechanical, others occasionally unintelligible, and even of the best of them sound like an immigrant with a noticeable accent. These should not be taken as limiting factors overall, since an android with a mechanical, occasionally unintelligible and highly accented patois has a certain charm. Your android would certainly be, in any case (and again, no pun intended), a conversation piece.

The reason phonemic approaches don't give us good results as simply as we might anticipate is that the specific sound of a phoneme is flavored not only by the inflections of normal speech but by the values of the phonemes both preceding and succeeding it.

The CT-1 Speech Synthesizer from Computalker Consultants (821 Pacific Street #4, Santa Monica, CA 90405), which uses a phonemic approach, sets target values for phonemes and charts smoothed parabolic paths between values for adjacent phonemes before actually pronouncing them.

Various versions of the CT-1 are priced at this writing between $425 and $600, with software and design aids available at extra cost. The reason for this is not so much the cost of the hardware in the speech synthesizer as the cost of adjusting it, fine tuning it, critically trimming it in order to make the various oscillators and filters (while digitally controlled, the speech generators are analog) sound like natural speech. Still, it's one of the best values available in a generalized vocabulary synthesizer.

Like all phonemic synthesizers, there is no inherent vocabulary to the CT-1. It accepts data inputs that sequentially drive the phonemes. It's up to our software and patience and skill to develop words based on combinations of phonemes.

Actually, I'm being unfair. The phonetic mode is only one of two modes available with the Computalker CT-1. The software package included with the CT-1 (which is designed for use on a host microprocessor system bus) converts ASCII character strings into phonetic text strings. For example, to say hello, enter HHEHLOW.

The other mode is a direct parameter control mode, in which the following nine speech parameters are controlled directly by data: voicing amplitude, voicing frequency, formant 1 (about 500 hertz) frequency, formant 2 (about 1,500 hertz) frequency, formant 3 (about 2,500 hertz) frequency, aspiration amplitude, frication amplitude, frication frequency, and nasal amplitude. A tenth data bit switches audio on and off.

Speech in this latter mode is *very* natural sounding. Unfortunately, a great deal of computer and hand work is needed in analyzing spoken words before the data codes for them can be identified. Computalker charges about $50 a second to extract parameter data from your tape recording.

They also report some progress in developing an all-digital board. But before I misquote them, why don't you write them for information if you want to know more. They have boards that are compatible with the S-100, Apple, and TRS-80 buses (maybe more by now), and a very informative literature package is available.

Votrax® (A Division of Federal Screw Works, 500 Stephenson Highway, Troy MI 48084) offers one of the best phonemic synthesizers available for under $4,000 (unit price, as of this writing)—their VS-6 Electronic Voice System. The unit measures about 11 × 10 × 3 inches and operates on ac line power. Prices start at about $3,500–$3,600.

The eight-bit command word presented to the unit's parallel input is divided into a six-bit selection code, which selects from among 61 phonemes, plus pauses and control functions; and a two-bit level-of-inflection code, which selects one of four pitches.

The VS-6, like the Computalker, links adjacent phonemes together smoothly through dynamic tracking. As a result, the specific output sound as a result of any given input code is varied by the context in which it appears.

The VS-6 contains no vocabulary of its own but rather pronounces sounds on the basis of data appearing at its input. Thus, it is an electronic simulation (or model) of the human vocal system. This voicebox operates under the instruction of an external "brain." Votrax reports one customer with a developed vocabulary of over 300,000 words in storage on the computer that drives its voice system.

For the Votrax VS-6 and other synthesizers based on the principle of *phonemic concatenation* (the speaking out of linked phonemes), vocabulary is virtually unlimited, but the quality of speech depends on the skill of the programmer.

The use of phonemic coding also allows a minimum of information (data) to address any given word. Well, almost minimum. Let's shift back to an overview.

In terms of data efficiency—in other words, how compact a data stream is required to output any given word or phrase—the worst case is full analog-to-digital conversion. This requires a data bandwidth equivalent to twice the frequency bandwidth of the speech band. Since the speech band must include such high frequency noise factors as sibilants and aspirants and fricatives, we must consider, in the worst case, a 10-kilohertz bandwidth, meaning a 20-kilobaud data rate. A minimum of eight bits permits realistic amplitude reproduction. Assuming an eight-bit byte, this means 20K bytes of memory for every second of speech.

If we can fit 2 words into a second of speech memory (this is a very realistic figure and a design center for most such speech storage approaches) a 240K disk can only store 20 words. What's more, this is a fixed vocabulary. On the other hand, since it plays back an essentially recorded (digitally recorded) voice, programming is easy and voice quality is excellent.

A number of shortcuts and compromises in voice quality permit increasing this efficiency from two to ten times. With changes in the coding approach, such as *continuously variable slope delta modulation* (in which serial data *go-up* and *go-down* instructions replace amplitude information—Motorola, for one, has integrated circuit CVSD chips and information available), data may be compacted again.

There are several examples of these data conversion approaches.

Steve Ciarcia has an excellent discussion of this approach in the June 1978 *Byte*, "Talk to Me!—Add a Voice to Your Computer for $35." He includes both a schematic and an 8080 assembler program. (For further reading on speech synthesis, see Appendix.)

Voicetek (P.O. Box 388, Galeta, CA 93017) is marketing Cognivox ($149 assembled as of this writing), a voice recording and recognition-scoring board that is compatible with the Z-80. It requires about 1½K bytes per second of speech. In the manufacturer's recommended configuration, it can recognize and reproduce a user-selected 16-word vocabulary. This fixed vocabulary is interesting but probably limiting.

Mountain Hardware (300 Harvey West Boulevard, Santa Cruz, CA 95060) offers SuperTalker, a speech digitizer/playback interface board designed for use with Apple (trademark of Apple Computer, Inc.). Four coding rates from ½K to 4K bytes per second of speech are selectable, and a variety of software is available. Steve North gave SuperTalker an excellent review in the October 1979 *Creative Computing*. It's possible, if not likely, that this board could be modified for on-board application in an android.

By comparison, the phonemic approach only needs 100–200 bits per second of speech, at eight bits per byte, for continuous speech. Overall, words tend to have as many phonemes as they do letters, an excellent rule of thumb for approximating data storage requirements.

Of course, the phonemic approach requires two levels of data handling. In the first, phoneme combinations for each vocabulary word must be partitioned into memory (which are the sequences of command words necessary to speak each word). In the second, which is optional, specific phrase sequence subroutines can permit speech output without extensive attention from the controlling processor(s).

Table 18-1 *Vocabulary of TSI Calculator Speech Synthesis Module S16001-A and S2A Mini Circuit Board.*

oh	percent
one	low
two	over
three	root
four	em (m)
five	times
six	point
seven	overflow
eight	minus
nine	plus
times-minus	clear
equals	swap

There is yet another level of efficiency, in which a large or small fixed vocabulary is addressed by only one or two bytes of data. And there are two excellent sources for hardware: TSI (Telesensory Systems, Inc., 3408 Hillview Avenue, P.O. Box 10099, Palo Alto, CA 94304) and Texas Instruments (P.O. Box 1444, M/S 7784, Houston, TX 77001).

TSI offers a number of small printed circuit modules with fixed vocabularies at excellent prices. With word-address inputs (one byte!) and a start command, these boards read ROM-stored data out through an analog converter/synthesizer and amplifier. The output is a male voice described by TSI as "clear, highly intelligible."

TSI currently offers three basic vocabularies, a 24-word calculator vocabulary, a 64-word "standard" vocabulary (useful for talking calculators and meters, for example) and a 64-word "ASCII" vocabulary. At this writing, unit prices range from $95 to $179. There is an option that permits you to order custom vocabularies for about $200 a word (or less) plus ROM mask or PROM programming charge (see Tables 18-1 through 18-3).

There also are two projected TSI products that may be of interest. One is a 5- by 6-inch board with about 130 words for about $625 (more for custom vocabulary) using *linear predictive coding*, which we'll look at in a minute. The other is an unlimited vocabulary system capable of accepting text, phoneme, or speech parameter (formant, etc., specifications) data as input.

TSI uses a technique called *linear predictive coding* to store speech code and decode data back into speech. Texas Instruments also makes extensive use of linear predictive coding, or LPC. If you've heard a Speak & Spell™, you know how natural a speaking voice this technique can produce.

Texas Instruments provides this definition of LPC: "the modeling of a speech waveform as the output of an all-pole recursive filter excited by either 'pseudo-random' noise or periodic pulses." In the TI realization of LPC, a ten-stage digital lattice filter (which responds to digital inputs in a way directly analogous to the way our brain controls the shapes of our tongues, lips, mouths, and air passages as we speak) alters two basic kinds of analog inputs to eventually produce speech.

These two kinds of analog signals are *voiced* (say "ahhhh") and *unvoiced* (things you can sound out without breathing out, like k-p-t-f).

Table 18-2 *TSI Model S2B 64-word "standard" vocabulary.*

zero	one	two
three	four	five
six	seven	eight
nine	ten	eleven
twelve	thirteen	fourteen
fifteen	sixteen	seventeen
eighteen	nineteen	twenty
thirty	forty	fifty
sixty	seventy	eighty
ninety	hundred	thousand
plus	minus	times
over	equals	point
overflow	clear	percent
and	seconds	degrees
dollars	cents	pounds
ounces	total	please
feet	meters	centimeters
volts	ohms	amps
hertz	dc	ac
down	up	go
stop	tone (low)	tone (high)
oh		

The digital bit stream is decoded into parameters, which are smoothed (so adjacent sounds are nearer each other) before being applied as controlling data for the lattice filter(s); and into energy and pitch data for the excitation sources (a voicing source and an unvoicing source).

The output of the lattice filter is converted from digital to analog, amplified, and outputted.

It takes a large computer and a lot of time to produce the proper data stream to produce a recognizable word with LPC—so don't even think of trying it yourself, even though decoding is easier than ever. Why so easy? TI put it on a chip.

The TMC0280 from Texas Instruments is a remarkable LSI IC that performs the complete data stream into analog speech signal conversion.

A lot of work went into the development of TMC0280 Speech Synthesizer IC. In April 1978, at a Speech Synthesis Technology Seminar, TI presented a briefing on its design goals in developing this IC:

- High quality speech for single word recognition
- Low data rate
- Low system cost, i.e., minimum external components
- Utilize lowest cost process (metal gate PMOS)
- Internal data processing flexibility to accommodate data rate modifications
- Simple external interface

Table 18-3 *TSI Model S2C 64-word "ASCII" vocabulary.*

space	x-point	quote
number	dollars	percent
and	apostrophe	left paren
right paren	star	plus
comma	minus	point
slash	zero	one
two	three	four
five	six	seven
eight	nine	colon
semicolon	less than	equals
greater than	mark	at
a	b	c
d	e	f
g	h	i
j	k	l
m	n	o
p	q	r
s	t	u
v	w	x
y	z	lower case
tone	upper case	up arrow
control		

Using tenth-order linear predictive coding, which provides a mathematical model for the human voice, this chip performs arithmetic manipulations to control ten stages of digital filtering. (In conjunction with other ICs in a 3-chip kernel system, a number of devices have been marketed by TI capable of excellent reproduction of fixed, precoded vocabularies. These include Speak & Spell™, a language translator, and a board-level module for computer voice output.)

Sounds perfect for an android, doesn't it? And the news gets a little better.

In November 1979, TI issued press release SC-233, "Speech Synthesizer Module Introduced by Texas Instruments," announcing 1980 production of the TM990/306 single board speech synthesizer module, designed for use with TI's TM990 microcomputer.

The features of the TM990/306 board are remarkable. It offers a fixed vocabulary of 179 words. The inputs and outputs are TTL-compatible. Its built-in event or interval timer can perform as a real time clock. While designed for use with TM990 CPUs, its external interface permits its use with virtually any processor, either in a polled-status or interrupt-driven mode. An amplifier on the board provides $2\frac{1}{2}$-watts drive to an 8-ohm speaker; a preamplifier output also is available. All of these features are on a $7\frac{1}{2} \times$ 11-inch PCB. At this writing, the unit price is $1,280.

The board has been designed for high intelligibility even in noisy environments. It produces excellent sound and speech quality, albeit the inflection is somewhat flat.

Power requirements are 5 VDC and $\pm$ 12 VDC. Two jumpers on the board select one of four possible interrupt levels; in its interrupt-driven or polled-status modes communications are parallel. On the TM990, this is memory-mapped I/O. On other host CPUs, an external D-type latch is required to hold the speech commands (the addresses of the words to be spoken) for interrupt-driven operation, the fastest mode. An interrupt occurs once a given spoken word has completed, and the host CPU can determine whether or not to issue a command to speak another.

Currently, the module requires about a tenth of a second between words (TI hopes to improve this), compared to about 2 milliseconds between words when we speak.

One way for us to improve this is through the use of an analog "bucket brigade" audio delay IC—available at your local Radio Shack for under $15. By letting the host speech CPU control its input and output clock rates (or simply inhibiting or deinhibiting the clock), we can clock speech data in, inhibit the input clock during pauses, and jam words together before they're outputted. This is especially helpful if we try to expand the resident vocabulary through permutations.

Huh?

Ham radio operators know what I'm talking about—it's a trick that originated with the old Morse Code telegraphers. They learned to abbreviate a call for attention, "seek you," to the letters "CQ."

Add the letters D or N to the word CREASE and you have approximations of INCREASE and DECREASE. And so on.

Before we leave the subject of speech synthesis, examine Tables 18-4 and 18-5 for an idea of how the base vocabulary of TI's board—which is typical, you'll find, of vocabularies of 100 to 200 words available from other speech synthesis sources—can be greatly expanded through homonyms (words that sound like other words), combinations, permutations, and puns.

Table 18-4 *First-order expansion of TI Model TM990/306 vocabulary through homonyms and simple combinations.*

a*	bee = b
abort*	beeline = b + line
abortive = abort + of	before = b + four
adjust*	beget = b + get
affects = f + x	behold = b + hold
alert*	below = b + low
all*	benign = b + nine
already = all + ready	beset = b + set
amps*	button*
and*	buttonholed = button + hold
annex = n + x	
any = n + e	c*
are = r	cagey or cagy = k + g
asinine = s + n + nine	calibrate*
ate = eight	call*
automatic*	cancel*
	canine = k + nine
b*	carpenter*
back*	carpentry = carpenter + e
backfire = back + fire	check*
backlight = back + light	clock*
backstop = back + stop	close*
backward = back + ward	closes = close + is
be = b	clothes = close
beady = b + d	clothesline = close + line

Note: Standard vocabulary words are marked with "*."

Table 18-4 *Continued.*

college = call + h
control*
crane*
crease*
creases = crease + is
cue = q
cycle*

d*
daisy = days + e
danger*
dangerous = danger + s
days*
daze = days
decrease = d + crease
decreases = d + crease + is
defeat = d + feet
degrees*
delight = d + light
device*
devices = device + is
direction*
display*
divisive = device + of
do you want to = two + u +
   one + two
door*
dory = door + e
down*
downturn = down + turn
downward = down + ward
downy = down + e

e*
east*
easter eggs = east + r + x
eastern = east + turn
eastnortheast = east + north
   + east
eastnortheastern = east
   + north + east + turn

eastnortheastward = east
   + north + east + ward
eastsoutheast = east + south
   + east
eastsoutheastern = east
   + south + east + turn
eastsoutheastward = east
   + south + east + ward
eastward = east + ward
easy = e + z
eaten = e + ten
effects = f + x
eggs = x
ego = e + go
eight*
eighty = eight + e
electrician*
eleven*
empty = m + t
enemy = n + m + e
energy = n + r + g
engage = n + gage
engages = n + gage + is
enlighten = n + light + n
enter*
entry = enter + e
envy = n + v
eon = e + on
equal*
ever = f + r
evergreen = f + r + green
examine = x + m + n
exit*
eye opener = i + open + r

f*
fail*
failsafe = fail + safe
farad*
fast*
fastback = fast + back
faster = fast + r

Table 18-4 *Continued.*

feet*
fire*
firelight = fire + light
firepower = fire + power
fiery = fire + e
five*
flow*
for = four
foray = four + a
fore = four
foreclose = four + close
forego = four + go
foreign = four + n
foreman*
forget = four + get
forty = four + d
forward = four + ward
four*
frequency*
from*
frumpy = from + p

g*
gage*
gate*
gauge = gage
gauges = gage + is
get*
ghetto = get + o
go*
goal line = go + line
goatee = go + t
green

h*
henry*
hertz*
hi = high
high*
highlight = high + light
hold*

holdover = hold + over
holdup = hold + up
hours*
hundred*
hurts = hertz

i*
ideal = i + d + l
if = f
imposition = m + position
impound = m + pound
impressive = m + press + of
in = n
inch*
inches = inch + is
increase = n + crease
increases = n + crease + is
indirection = n + direction
indoor = n + door
inequal = n + equal
infrequency = n + frequency
initialize*
initializes = initialize + is
inn = n
innate = n + eight
input = n + put
inspector*
inundate = n + and + eight
inward = n + ward
is*
isn't = is + n

j*
jog*

k*

l*
left*
lefty = left + e
light*

Table 18-4 *Continued.*

lighten = light + n
lightweight = light + wait
line*
low*
lowdown = low + down

m*
machine*
machinery = machine + r + e
maker*
manual*
manually = manual + e
measure*
mega*
meter*
micro*
mill*
milli*
minus*
minuses = minus + is
minutes*
motor*
move*
movie = move + e

n*
nine*
ninety = nine + d
north*
northeast = north + east
northeastern = north + east
    + turn
northeastward = north + east
    + ward
northnortheast = north + north
    + east
northnortheastern = north
    + north + east + turn
northnortheastward = north
    + north + east + ward
northnorthwest = north + north
    + west

northnorthwestern = north
    + north + west + turn
northnorthwestward = north
    + north + west + ward
northward = north + ward
northwest = north + west
northwestern = north + west
    + turn
northwestward = north + west
    + ward
number*

o*
o'clock = o + clock
odor = o + door
of*
off*
office = off + s
offset = off + set
often = off + n
ohms*
oh-oh = o + o
okay = o + k
omega = o + mega
on*
one*
onesy = one + z
onset = on + set
onto = on + two
onward = on + ward
open*
opener = open + r
operator*
orange = o + range
our = r
ours = hours
out*
outdoor = out + door
outline = out + line
outnumber = out + number
output = out + put

Table 18-4 *Continued.*

outset = out + set
outward = out + ward
oven = of + n
over*
overall = over + all
overflow = over + flow
overpower = over + power
overtime = over + time
overturn = over + turn
overweight = over + wait

p*
pass*
passed*
passes = pass + is
passive = pass + of
password = pass + ward
past = passed
pastel = passed + l
pastime = pass + time
pastor = passed + r
peon = p + on
people = p + pull
percent*
percentage = percent + h
pico*
plus*
plusses = plus + is
point*
pointy = point + e
policed = pull + east
position*
pound*
power*
press*
presses = press + is
pressure*
priority*
probe*
processing*
pull*

pulley = pull + e
push*
pushes = push + is
pushy = push + e
put*
putdown = put + down

q*

r*
range*
ranger = range + r
ranges = range + is
read (past tense) = red
ready*
red*
repair*
repeat*
repeater = repeat + r
replace*
replaces = replace + s
right*
righty = right + e

s*
safe*
safety = safe + t
sea = c
sealine = c + line
seaward = c + ward
seconds*
see = c
seedy = c + d
seek you = c + q
set*
setpoint = set + point
settee = set + t
seven*
seventy = seven + t
shut*
shutdown = shut + down

Table 18-4 *Continued.*

shuteye = shut + i
shutter = shut + r
six*
sixty = six + t
slow*
smoke*
smoky = smoke + e
south*
southeast = south + east
southeastern = south + east
  + turn
southeastward = south + east
  + ward
southsoutheast = south + south
  + east
southsoutheastern = south
  + south + east + turn
southsoutheastward = south
  + south + east + ward
southsouthwest = south + south
  + west
southsouthwestern = south
  + south + west + turn
southsouthwestward = south
  + south + west + ward
southward = south + ward
southwest = south + west
southwestern = south + west
  + turn
southwestward = south + west
  + ward
speed*
speedy = speed + e
start*
stop*
stoplight = stop + light
switch*
switches = switch + is

t*
tango = ten + go

tee = t
temperature*
ten*
tender = ten + door
tennis = ten + s
test*
testy = test + e
the*
thousand*
three*
time*
timer*
to = two
tool*
toolmaker = tool + maker
turn*
turnip = turn + up
twelve*
two*

u*
under*
underflow = under + flow
undergo = under + go
underline = under + line
underrate = under + eight
underweight = under + wait
underwrite = under + right
unit*
unity = unit + e
up*
uphold = up + hold
upon = up + on
upright = up + right
upturn = up + turn
upward = up + ward

v*
valve *
volts*

Table 18-4 *Continued.*

w*
wait*
waiter = wait + r
ward*
warden = ward + on
watts*
weight = wait
welder*
west*
western = west + turn
westnorthwest = west + north
  + west
westnorthwestern = west
  + north + west + turn
westnorthwestward = west
  + north + west + ward
westsouthwest = west + south
  + west

westsouthwestern = west +
  south + west + turn
westsouthwestward = west
  + south + west + ward
westward = west + ward
why = y
word = ward
wordy = ward + e
write = right
writer = right + r
x*
y*
yellow*
you = u
z*
zero*
ziti = z + d

Table 18-5 *More words that can be formed from the base vocabulary, with pronunciation approximate.*

afraid = f + red
ago = a + go
ahold = a + hold
aisle = i + l
align = a + line
andes = and + days
android = and + door + right
antsy = and + c
attorney = a + turn + e
auntie = and + t
await = a + wait

backhand = back + and
backup = back + up
beast = b + east
be seein' you = b + c + n + u

breakfast = back + fast
buttin' in = button + n

desire = days + i + r
disengage = days + n + gage

effendi = f + n + d
entertain = enter + ten
entreaty = enter + e + t
entrophy = enter + v
entropy = enter + p
estate = east + eight
eternal = e + turn + all
excusee = x + q + z
exist = x + east

Table 18-5 *Continued.*

failure = fail + r
fascinate = fast + n + eight
fascinatin' = fast + n + eight
  + n
fasten = fast + n
fastest = fast + s
floor = flow + r
force it = four + set
forehand = four + and
forehead = four + red
formal = four + mill
forum = four + m
furlough = four + low

gem = g + m

gin = g + n
goal = go + l
gopher = go + four
go wait = go + eight
gringo = green + go

handle = and + l
have = f
haymaker = a + maker
hazy = a + z
he = e
hello = l + o
he stays = east + days
hey you = a + u
his = is
holdin' = hold + n
honor = on + r
hygiene = high + g + n

icy = i + c
interest = enter + s
interestin' = enter + s + ten
intern = n + turn

jam = j + m

kilo = k + low

lightest = light + test
liner = line + r
lower = low + r

may the force be with you =
  mega + four + s + b + wait
  + u
measurin' = measure + n
miller = mill + r
million = milli + n

numberin' = number + n

openin' = open + n

peachy = p + g
piccolo = pico + low
pointer = point + r
pressin' = press + n
pullin' = pull + n
pushin' = push + n

sedate = set + eight
sequel = c + equal
setter = set + r
settle = set + l
shake it up = j + get + up
shuttle = shut + l

tempest in a teapot = t + m
  + passed + n + a + t + put
tenor = ten + r
tester = test + r
thank you = ten + q
these = the + is
tobacco = two + back + o
today's = two + days
took you = two + q
twist = two + east

Table 18-5 *Continued.*

---

waistcoat = west + got
waitin' = wait + n   yes = e + s
want to = one + two

  You can work out what you *want* to say *when* from this expanded vocabulary and some good guesses at the conditions your android is likely to have to encounter—and talk about. It won't take much memory to store word addresses for these phrases in memory partitions, access them on command of a controller, and soon have your beastie talking as intelligently as any rock lyric, parakeet, or situation comedy. With a little work, you may even work your way up to politician.

19
Verbal Literacy

Speech recognition is a relatively undeveloped technology, despite millions of dollars poured into it each year by such giants as IBM, Bell Telephone, and the Department of Defense.

We know what we need to accomplish. We know how to accomplish large portions of the task. We don't quite know how to put the whole thing together and make it work.

For the android design task, let's approach the problem methodically. First, we'll try to define it. Second, we'll examine the problems and opportunities involved. Third, we'll review the hardware. Fourth, we'll throw in a couple of hints and insights into experiments you can try in an effort to improve the state of the art—or at least, make it more affordable at a given level of inaccuracy.

In the overview, all speech recognition systems try to perform some sort of match between an input signal including speech and a stored pattern corresponding to a specific word, providing a unique output for each word in its vocabulary it recognizes (see Fig. 19-1).

Ideally, it wouldn't make any difference who is talking, how much background noise is present, what dialect is being spoken, or whether or not the words are being run together. As usual, the ideal isn't real.

Consider that the android may or may not be facing the speaker, has no control over background noise, and may or may not be familiar with the speaker. There's no assurance that the speaker will enunciate properly, not have a head cold, and not speak with an accent. There also is no assurance that a given word will be in the android vocabulary, or might not be misinterpreted.

We can make some assumptions to ease the design task and some compromises to ease it even more.

First, there is only one word (in English) that the android should be able to recognize regardless of the speaker: "no!" Pet owners fully understand the significance of this decision. Second, the android can always be made to speak a word or two—"repeat", for example—to prompt a human to refine his or her input. Similarly, the android could

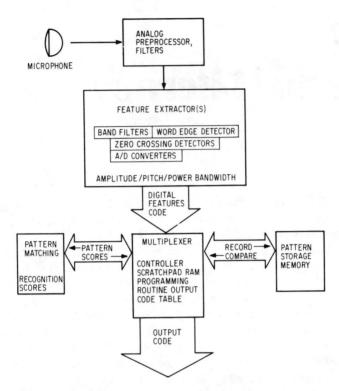

Fig. 19-1 *Simplified block diagram of speech recognition circuitry (generalized).*

say something like "Wait. Stop and wait from word to word." This would help make sure that words are separated, an important factor in simplifying the speech recognition problem.

A "speech finder" can compare binaural (or trinaural) amplitude and phase information from microphones positioned at several points around the head and be used to point the head or a directional mike directly at the speaker—in much the same way that a dog, for example, will rotate his ear shells to try to improve the audio quality of the sounds he hears.

The hardest decision is to limit the number of people for whom a large spoken vocabulary is recognized to one or two—at most four or five. So you or you and the spouse or you and the spouse and the kids can be recognized most of the time, and others may or may not be recognized ever.

Of course, you can multiply the memory needed and the processing time needed to recognize a particular word by the number of people the machine has to recognize. Or multiply the memory by the number and process all the voices in parallel with separate hardware.

There are some other options available, too, that we'll be seeing in our review of hardware and techniques.

Remember that we've said that all speech recognition schemes involve some attempt to match what they hear to what they've been taught a word should "sound" like, then identify the word. (For further reading on speech recognition, see Appendix.)

For our purposes, let's restrict our discussion to those versions of this process in which a reasonably small microprocessor system might be involved, albeit at some near future state of the art.

This means that we'll be dealing, in every case, with some eventual conversion of speech to a digital electronic signal. Straightforward analog-to-digital conversion in full audio fidelity with 12-bit amplitude values means a number of nasty things. One is a megabit memory for every 4 seconds of speech (8–10 words). Another is that every change in the specifics of pitch, ambient noise, pronunciation, and so on will change the coding for a word to some extent. Another is that it requires extremely fast circuitry to perform calculations against this large a chunk of input data in real time.

We don't need high fidelity, for one thing. You do a perfectly adequate job yourself of understanding speech you hear over a telephone, even though the highest frequency the telephone circuitry for a normal phone will pass is a third, more or less, of the 10-kilohertz bandwidth of the voice. Apparently, we can excise large amounts of data, if carefully selected, and still end up with recognizable speech. And there's an important key in that.

We can assume as an indicator, if not a measure, that our own ability to hear and understand speech that's undergone a great deal of processing corresponds to the extent to which a digital system succeeds with the same signal.

It turns out that amplitude information is all but useless. It's useful for telling when speech is or isn't present and helps locate the ends of words, but little else.

Most of the information we require is available by counting the number of zero crossings the speech waveform makes within a specified passband (the number of times the waveform crosses zero).

On the other hand, since we're talking about passbands, there may be a way to use some minimal amount of well-massaged amplitude information to evolve a unique data set for any word. Consider, for example, a series of bandpass filters, each sampling a given slice of speech spectrum. In mathematical terms, we can perform a first-order integration of amplitude versus time. In signal terms, we can detect its envelope. In hardware terms, we can rectify the narrowband signal coming out of each filter, smooth it with a capacitor, and arrange a decay time such that 50 to 150 samples per second give us all the envelope data we need. In human speech, envelopes vary slowly.

Envelopes alone are inexact indicators. But envelopes in conjunction with zero crossing analysis provide a wealth of information at a reasonably small data rate.

Because of the importance of sibilants, fricatives, and such in speech, at least one of the filters should be a simple high-pass filter at 3 to 4 kilohertz.

In human speech, there are frequency bands (called formant frequencies) in which energy is most likely to appear. The first of these (formant F1) spans 200 to 1,000 hertz. The second, F2, overlaps a bit and spans 800 to 3,500 hertz. The specific frequency (of zero crossings) within these bands does pretty well at identifying most vowel sounds, for example.

Usually, we are concerned with three formants, centered around 500, 1,500, and 2,500 hertz. We can select passbands that sample these center frequencies and selected adjacent sub-bands within each formant band, looking both for some rudimentary envelope value and for a zero crossing count in each.

Many elements of human hearing are essentially logarithmic. You may notice, for example, that a log taper volume control potentiometer seems to increase apparent "loudness" (a subjective phenomenon) linearly through its rotation.

So any analog-to-digital conversion should be logarithmically compressed. The *companding A/D converter* is a relatively new, relatively inexpensive IC capable of performing this conversion.

Also, logarithmic spacing of sampling sub-bands seems to give better results than linear spacing (versus frequency). And at higher frequencies, broader passbands are permissible; this is analogous to human perception of speech data.

In theory, a perfect data extractor (an exact analog to human hearing) would provide about 30 key features each second. Hardware still is nowhere near this capability for low data rate.

There is no perfect speech recognition hardware extant, in terms of our android's needs. Before we take a look at the imperfect hardware that is available, let's take a pseudorandom tour through some of the most salient information available about the speech recognition problem.

The ideal mechanism would respond to any speaker, any dialect, any casual speech, even run-together words, all with perfect understanding.

The minimum elements of a system include a microphone, some filters, some analog-to-digital technique or mechanism, and a microcomputer.

In simple systems, the data that comes out of the "front end" processors is compared to stored whole-word or whole-phrase data. In more complex systems, it is first matched against phonemes; phoneme strings are then compared against stored vocabulary. (The most elemental sound in language, there are between 40 and 100 phonemes, depending mostly on the universe of languages and dialects they attempt to encompass and on who's counting.)

Voiced sounds, like vowels, are produced by glottal excitations; glottal excitations carry little information in terms of word recognition or meaning. Nevertheless, the formants (which are the specific frequencies to which the glottal pulses are modified by the vocal tract) carry most information needed for word recognition and meaning. An excellent example of this is whispered speech, which is full of intelligence even in the absence of glottal excitation. So recognizing speech has nothing to do with the speaker's tone or pitch of voice.

Systems that are capable of understanding connected speech (no pause between words), as a rule of thumb, cost about six times as much as systems that can recognize isolated—or at least separated—words.

We haven't gone into any detail on how pattern matching is accomplished, and it isn't an area likely to provide absolute best choices—either in the microscopic or macroscopic sense.

The best systems seem to rely on either a process of elimination or a system of scoring each respective choice in order to determine to which vocabulary entry an input most closely corresponds.

One system adds an input data word to the complement of each vocabulary data word and looks for high scores. Another subtracts the input word from the vocabulary word and looks for a low score. Others multiply. Still others provide complex "distance" scores to determine how close or how far an input is from each vocabulary word: its output is not a single code but a schedule of probabilities for the most likely matches.

Still other systems provide contextual or syntaxical evaluations, which permit the schedules of probabilities to be weighted in favor of classes of words that *should* occur at any point.

Both parallel and serial schemes abound for examining large classes of objects, such as speech recognition vocabularies. Content Addressable Memory (CAM, a parallel scheme) and hashing (a serial scheme) are both worthy of further research by the reader.

Actually, the manufacturers of speech recognition hardware usually supply some mechanism—hardware, software, or both—by which the whole recognition task may be accomplished. This sounds like a cue for our long promised hardware review.

One of the better speech recognizers is named Mike™, a product of Centigram Corporation (155A Moffett Park Drive, Suite 108, Sunnyvale, CA 94086). The version you'll most likely be interested in is Model 3700—electronics only, requiring a power supply (12 VDC at 150 milliamperes, 5 VDC at 1.3 milliamperes, and −12 VDC at 60 milliamperes) plus a microphone and keyboard during programming. It also requires a Model 0010 I/O interface. At this writing, suggested U.S. resale for these items is $1,750 and $350, respectively. An additional $350 buys an option—Model 0041—which increases Mike's storage capability from two 16-word vocabularies to twelve. Both options are on printed circuit cards designed to ride piggyback on the main 10-by-10-inch PCB.

Mike uses digital techniques in place of many of the analog processing steps we've been talking about. The system is so innovative, the following description is quoted, with permission, from Centigram's brochure.

*How Mike™ Operates**

Mike learns and recognizes patterns derived from spectrum analysis data. When learning a word, Mike stores patterns in memory for future reference. When attempting to recognize a word, Mike compares the incoming pattern to each reference pattern and generates a set of "closeness to fit" scores. Above a certain threshold, the highest score is taken to indicate successful recognition.

The spectrum analysis is performed every 25 milliseconds to measure the energy in 16 logarithmically spaced frequency bands over the 300 Hz to 3,000 Hz range. Mike's approach to this analysis is unique. The data to be analyzed is spun past a single filter 16 times, each time at a different frequency, so that the frequency of interest matches the center frequency of the filter. This is in contrast to the conventional approach, which involves using 16 individually tuned filters operating in parallel.

The spectrum analysis data is digitized and passed to the word-framing process. When a sufficient level of spectral activity is detected, the beginning of a word is marked. When this activity falls below a threshold, the end of that word is marked. Since Mike is an isolated-word recognition device, a silent interval of approximately 100 milliseconds is required at the end of a word to frame it adequately.

---

* Copyright © 1979 by Centigram Corporation. Reprinted by permission of Centigram Corporation.

Noise-canceling and time-base normalization are integral parts of the word-framing process. During silent intervals, constant (ambient) noise is measured; during word framing, this constant noise signal is subtracted from the input signal. When a word or segment of sound has been isolated, it is normalized to a fixed time duration to compensate for different speaking rates.

The pattern generation process further operates on the framed word to extract features of interest and to reduce it to a string of approximately 240 bits. The pattern is then generated using a proprietary mapping algorithm.

In training Mike, patterns are logically OR-ed with the patterns of previous repetitions of the word being learned. Typically, two or three repetitions of each vocabulary word suffice for reliable recognition. When Mike is attempting to recognize, patterns are compared by AND-ing them in turn with each of the previously learned reference patterns. The matching ONEs are tallied to form a set of scores for each comparison.

Mike recognizes a word if its score is both above a threshold and greater than the next highest score by a prescribed increment. A code indicating the identity of the recognized pattern is transmitted to a host device. If a word is framed but does not meet the recognition criteria, a no-recognition code is transmitted.

Centigram's recognition approach is patented in the United States (Patent No. 4,087,630), and patents have been applied for in 15 other countries.

One of the newest manufacturers of speech recognition hardware at this writing is Voicetek (P.O. Box 388, Galeta, CA 93017). While company president Bill Georgiou won't discuss details of how the Voicetek Cognivox™ works, he has forwarded information on what it does; some insight into his approach might or might not be gained through his article "Give an Ear to Your Computer" in the June 1978 *Byte*.

Cognivox is a voice input/output peripheral available in versions designed to be used with the Exidy Sorcerer™; with general Z-80 based systems, including S-100 systems; and an announced TRS-80 version. Including speaker, amplifier, and enclosure, it measures $6\frac{1}{4} \times 3\frac{3}{4} \times 2$ inches, consumes 4 milliwatts during speech recognition and 100 milliwatts when producing sound. The Z-80 interface is via two input port bits and one output port bit.

At this writing, suggested U.S. resale for the Cognivox is $149. assembled and tested, plus a 3 percent shipping and handling charge.

Its vocabulary is only 16 words. It requires about 1½ kilobytes of RAM for each second of speech and about 3K for its controlling program in BASIC. The Cognivox is not very good at discerning between similar-sounding words. It requires the user to speak clearly, to separate words, and to speak directly into the microphone—none of which can be assured for an android.

Still, if your Z-80 microprocessor and cash are not otherwise committed, Cognivox can provide an excellent introduction into the problems and opportunities speech recognition can present.

There are a number of manufacturers of highly capable, very expensive and highly accurate commercial speech recognition systems, most notably Centigram Corporation, Threshold Technology (1829 Underwood Boulevard, Delran, NJ 08075), Dialog Systems, Inc. (639 Massachusetts Avenue, Cambridge, MA 02139), Scope Electronics, Inc. (Reston, VA 22070), and Perception Technology, Inc. (Winchester, MA 01890). By and large, these companies are offering speech recognition systems for industrial applications and are not included here because, for various reasons (and with the exception of the Mike PCB version discussed previously), they are not easily applicable to the android design requirements.

One company that does very well in providing reasonable speech recognition hardware at reasonable prices is Heuristics, Inc. (1285 Hammerwood Avenue, Sunnyvale, CA 94086). An excellent example of the quality of product they manufacture is the SpeechLink™ 2000, suggested U.S. resale priced at $259 at this writing. It's designed to be used with the Apple® computer (and probably could be jockeyed around to work with other 6502 hardware) and provides a user-variable 64-word vocabulary.

SpeechLab®, the earlier series of Heuristics products, offers Apple and S-100 compatible boards with 32-word vocabularies (Models 20A and 20S, suggested U.S. resale $189) and a CMOS S-100 compatible 64-word vocabulary model (Model 50, $299). These offer ROM programming, relocatable under the control of the host system to any 2K address; an additional 4K of RAM is used by the program.

The Model 20 includes two bandpass filters with two amplitude bits, two zero crossing detectors, and a linear amplifier. The Model 50 includes three bandpass filters, one zero crossing detector, a linear amplifier, a compression amplifier, an analog-to-digital converter (6 bits), and a prompting beep generator.

Heuristics also offers the manuals to their products separately. Consult them for current price and availability information.

The choice of a microphone is an extremely important consideration. Noise-canceling characteristics, frequency response limited to the voice band, immunity to variations of response with temperature and humidity, and light weight are important considerations. Directionality is *probably* an important requirement; the builder will want to experiment with both the "speech finder" approach mentioned early in this chapter and with directionalizing geometries for the microphone housing—analogous to the ear shells and cones of humans and animals.

Knowles Electronics (3100 North Mannheim Road, Franklin Park, IL 60131) is a leading manufacturer of subminiature acoustic transducers for over a quarter of a century. Their microphones are used in hearing aids (appropriate!), telephone headsets, pagers, dictating machines, and elsewhere. Write them for information; while they sell primarily to OEMs (original equipment manufacturers), they may be willing to sell to individuals or clubs as well or may be able to suggest another way to purchase their fine products.

One last consideration (though far from a final one) is the determination of what words you wish to provide for in your android's speech recognition vocabulary. In addition to the word "no," you may want to consider a name for the android as well as a number of key words. Other human sounds, like laughter, might also deserve consideration.

Table 19-1 lists recommended vocabulary words for speech recognition and is included here as a starting point for your own determinations. These words and their homonyms are included, as well as categories of usage; while grammatical terms are indicated, let me emphasize that the categories correspond to actual usage more so than to their proper grammatical categories. After all, it would do us little good to program an android to accept only rigidly perfect language when we ourselves are probably sloppy users of the rules of grammar.

Table 19-1 *Recommended speech recognition vocabulary.*

Homonyms are listed in parentheses. In addition, words are coded as belonging to one of these eight categories:

1. *Imperative verb*
2. *Transitive verb, intransitive verb*
3. *Adverb, conditional*
4. *Assent*
5. *Denial*
6. *Noun, pronoun*
7. *Formalities*
8. *Adjective*

ADD (AD)   1,2,6,8
AFFIRMATIVE   4,8
AFTER   3

AGAIN   3
ALL (AWL)   3,6,8
ANIMAL   6,8

Table 19-1 *Continued.*

ARE (R) 2,6
ARM 2,6,8
ATTENTION 1,3,6

BACK 1,2,3,6,8
BAD 8
BAG 2,6,8
BATHROOM 6,8
BATTERIES 6
BEDROOM 6,8
BEGIN 1,2
BIG 3,8
BOOK 6,8
BOTTOM 6,8
BOX 2,6,8
BOY 6,8
BRING 1,2
BROKEN 8
BUTTON 2,6,8
BY (BUY, BYE) 2,3,6,7

CAR 6,8
CAT 6,8
CHAIR 6,8,
CHECK 2,4,6,8
CHEST 6,8
CIRCLE 2,6,8
CLOCK 2,6,8
CLOSE (CLOTHES) 1,2,3,6,8
CLOSE (as in not far) 3,8
COFFEE 6,8
COLD 3,5,6,8
COLUMN 6,8
COME 1,2
CONTROL 1,2,6,8

DANGER 6
DARK 3,6,8
DIFFERENT 3,5,8
DOG 6,8
DO (DEW) 1,2,4,6

DOOR 6,8
DOWN 1,3,6,8
DRAWER 6,8
DRESS 2,6,8
DRINK 2,6,8
DROP 1,2,6

EDGE 1,6,8
EIGHT (ATE) 2,6,8
END 1,2,6,8
ENOUGH 1,3,6,8
EQUAL 2,6,8
ESCAPE 1,2,6,8
EVEN 2,3,6,8
EYES 2,6,8

FAR 3,6,8
FAST 3,8
FAUCET 6,8
FETCH 1,2,6
FIND (FINED) 1,2,6
FINGER 3,6,8
FIRE 1,2,6,8
FIRST 3,8
FIVE 6,8
FIX 1,2,6,8
FOLLOW 1,2
FOOD 6,8
FOOT 6,8
FORWARD (FOREWARD) 1,2, 3,6,8
FOUR (FOR, FORE) 3,6,8
FROM 3
FRONT 3,6,8

GARAGE 2,6,8
GET 1,2
GIRL 6,8
GIVE 1,2,6
GLASS 6,8
GO 1,2

Table 19-1 *Continued.*

GONE 3	LOUD 3,8
GOOD 3,4,6,8	LOW 3,8
GRASP 1,2,6	
	MAGAZINE 6,8
HALL (HAUL) 1,2,6,8	MAKE 1,2
HAND 1,2,3,6,8	MAN 1,2,6,8
HE 6	ME 6
HEAD 1,2,6,8	MORE 3,6
HEAVY 3,6,8	MORNING (MOURNING) 6,8
HELLO 7	MOUTH 2,6,8
HELP 1,2,6	MOVE 1,2,6
HER 6,8	MY 6,8
HERE 1,3	
HIGH (HI) 3,7	NAME 1,2,6,8
HIS 6,8	NARROW 1,2,6,8
HIT 1,2,6,8	NEAR 3,8
HOLD 1,2,6	NEGATIVE 5,8
HOME ´2,3,6,8	NEWSPAPER 6,8
HOT 3,8	NICE 3,4,8
	NIGHT (KNIGHT) 6,8
I (AYE, EYE) 2,4,6,8	NINE (NEIN) 5,6,8
IF 3	NO (KNOW) 1,2,3,5,8
IN 3	NOT (KNOT) 2,3,5,6,8
IS 2	NOW 3
	NUMBER 1,2,6
JUST 3,8	
	OF 3,8
KITCHEN 6,8	OFF 1,3,8
KNOB 6	OKAY 1,2,3,4,6,8
	ON 1,3,4,8
LEAN 1,2,3,6,8	ONE 6,8
LEFT 3,6,8	ONLY 3,8
LEG 2,6,8	OPEN 1,2,3,8
LESS 3,6,8	OTHER 6,8
LIFT 1,2,6	OUR (HOUR) 6,8
LIGHT 1,2,3,6,8	OUT 1,3,8
LIQUID 6,8	OUTLET 6,8
LISTEN 1,2	OUTSIDE 3,6,8
LIVING ROOM 6,8	OVEN 6,8
LONG 3,8	
LOOK 1,2,6	PANTS 2,6,8

Table 19-1 *Continued.*

PAPER 1,2,6,8	SLOW 1,2,3,8
PERSON 6	SMALL 8
PIECE (PEACE) 1,2,6,7,8	SOCKET 1,2,6,8
PLATE 2,6,8	SOLID 3,6,8
PLEASE 3,7	SOME 3,6
PLUG 1,2,6,8	SQUARE 1,2,3,6,8
POINT 1,2,6,8	STAIRS (STARES) 2,6
POT 1,2,6,8	START 1,2,3,6
POWER 1,2,3,6,8	STOP 1,2,3,6
PROBLEM 6,8	STOVE 2,6,8
PULL 1,2,6	STRONG 3,8
PUSH 1,2,6	
PUT 1,2,6	TABLE 6,8
	TAKE 1,2,6
QUESTION 1,2,6	TALL 8
QUIET 1,2,6,8	TEA (TEE) 2,6,8
	TELEPHONE 1,2,6,8
RAMP 1,2,6	TELL 1,2
READ (REED) 1,2,6,8	THANK YOU 7
REFRIGERATOR 6,8	THAT 6,8
REPEAT 1,2,3,6,8	THE 6,8
REVERSE 1,2,3,5,6,8	THEN 3
RIGHT (WRITE) 1,2,3,4,6,8	THERE (THEIR) 3,6,8
ROOM 2,6,8	THESE 6,8
ROTATE 1,2	THING 6
ROW 1,2,6,8	THIS 6,8
RUN 1,2,3,6,8	THOSE 6,8
	THREE 6,8
SAME 3,8	THROUGH 3,8
SET 1,2,6,8	TILT 1,2,6
SEVEN 6,8	TIME 1,2,6,8
SHAPE 1,2,6	TIRED 3,8
SHE 6	TOP 1,2,6,8
SHIRT 6,8	TRUCK 1,2,6,8
SHOE 6,8	TURN (TERN) 1,2,6
SHORT 3,6,8	TV 6,8
SHOW 1,2,6,8	TWO (TO, TOO) 3,6,8
SIDE 1,2,3,6,8	
SIDEWALK 6,8	UNDERSTAND 1,2
SINGLE 1,2,8	UP 3
SIX 6,8	US 6

Table 19-1 *Continued.*

USE 1,2	WHOLE (HOLE) 6,8
	WIDE 8
WARNING 1,2,6,8	WOMAN 6,8
WATCH 1,2,6,8	WRONG 3,6,8
WATER 1,2,6,8	
WE (OUI, WEE) 3,4,6,8	YES 4
WEAK (WEEK) 6,8	YET 3
WHAT 6,8	YOU (EWE, U) 6
WHEN (WEN) 3,4,5,6	YOUR 6,8
WHICH (WITCH) 6,8	
WHO 6,8	ZERO 1,2,3,5,6,8

Note: Keys indicated include many ungrammatical but common uses of the indicated words. You may wish to select from this list for your own purposes or devise a list of your own.

Note that we haven't discussed the need to interpret strings of recognized words into actions. We'll discuss the principle, at least, when we take a closer look at how the various microprocessors are organized.

20
Brains

So far, the android we've built has almost all the qualifications of the Tin Woodsman of Oz. Now if he only had a brain . . .

If he only had a brain, he wouldn't be able to get much done, because one brain isn't enough to process all the raw information that sensory circuits provide and motor circuits require—at least not one microprocessor, no matter how brainy.

But we can copy from nature again and design a nested hierarchy of processors that work together to give the android as much of a mind as you care to invest in developing. And once the basic nuggets are in place, additional brainpower can be layered on. The more layers, the more of a mind the android will possess. And it doesn't matter when you decide to upgrade with another higher-level processor, which is about the most clear-cut case of mind over matter I've ever heard of.

In this chapter, we're going to look at how these processing layers should be organized, how they should communicate, and who's boss. We'll take a look at some of the options available in selecting microprocessors—and make no mistake, you'll be using dozens of them. And you'll find plenty of places to go to for more information.

But in this chapter—in fact, in this book—you won't find any programs, no fixed routines or subroutines, no software cast in cement, no specifics. Instead, you'll learn how to develop these for yourself, or at least the rudiments of them.

In the April 1978 issue of *Interface Age*, Northeastern Regional Editor Roger C. Garrett authored a very interesting article entitled "A Natural Approach to Artificial Intelligence." In a unique if somewhat simplistic approach to multiple processor organization in an android (or robot, as originally addressed), Roger suggested a distributed processing scheme with processing located at every key point of anatomy, whether sensory or motor, organized like the branches of an anatomical tree. He suggested modularizing the approach according to three rules:

1. Each processing module or center of intelligence must have control of all key points that are simpler than it is, according to the strict chain of command of the branch.

2. Each processing module or center of intelligence can send instructions and commands to all modules that are lower on the chain of command than it is.

3. Each processing module or center of intelligence can send information or requests for assistance up the chain of command to all modules that are less simple than it is, or at least more responsible.

The chain-of-command approach is a good one and in many cases can be adopted or adapted almost directly. There are enough cases where the "thigh bone's connected to the hip bone" approach is inadequate, however, that it's worthwhile to look a little farther before trying to model the organization of our android's mind (see Figs. 20-1 through 20-5).

With a little soul-searching introspection (okay, you got me, I'm being redundant), you'll realize that the defining factor in specifying a level of responsibility isn't anatomical in any sense but purely informational.

Here, we're going to depart a slight bit from traditional thinking in artificial intelligence; where traditionalists prefer to speak in terms of information and information processing, we can more quickly grasp the tasks facing our several processors if we think instead in terms of *knowledge*, the product of processed information. Each center of intelligence has both a need to know certain things and an ability to share its acquired knowledge (see Table 20-1).

The process of organizing these centers of intelligence (which may each require a single microprocessor or several) begins with assigning them tasks; that, in turn, begins with a complete listing of the tasks the android will ever have to accomplish, in the sense of knowledge assimilation and distribution. Whoa, slow down, we'd better get a little more basic.

Each sensor throughout the android is a source of some specific element of knowledge. For example, a microswitch will know when something has pressed its button enough to turn it on. Our ramera will know when light has depleted any specific location in its array. This is all crude, unprocessed knowledge (okay, information, either word is fine here).

Our eventual goal is to give these bits of knowledge the power to assume meaning, which means providing some mechanism for understanding. No, these are not terms often used with data processing, but they are not necessarily terms that need to be associated with high-level organisms, either.

Knowledge, meaning, and understanding in very crude forms exist in your refrigerator. The designer has designed in a small switch that closes when the door is open. That switch imparts the knowledge that the door is open. The switch is connected in a simple circuit that lights a light bulb. So the knowledge of the door being open means the

Table 20-1 *Centers of Intelligence.*

I. Overseer processor
  A. Receives instructions from humans
    1. Voice
    2. Keyboard
    3. Override (panic)
  B. Responds to alarm conditions
    1. Collision
    2. Entrapment
    3. Low energy
  C. Sets immediate operational goals
    1. Responds to instructions
    2. Responds to alarm conditions
    3. Maps immediate area
      a. Analyzes visual inputs
      b. Analyzes contact inputs
    4. Learns about new objects
      a. Learns to recognize by sight
      b. Learns to recognize by name
      c. Asks questions, may format answers
    5. Searches for and locates known objects, enhances map
    6. Learns new operations (e.g., vacuuming rug or pouring a cola)
    7. Performs subset of known operations

II. Intelligent peripherals
  A. VISUAL subsystem
    1. Obstacle recognition
    2. Free path recognition
    3. Object recognition
    4. Range calculation
    5. Mapping input
  B. DRIVE subsystem
    1. Collision, obstacle avoidance
    2. Mapping
    3. Navigation
    4. Drive motor control
  C. SPEECH RECOGNITION subsystem
  D. SPEECH SYNTHESIS subsystem
  E. Left, right MANIPULATOR control subsystem
  F. Future features

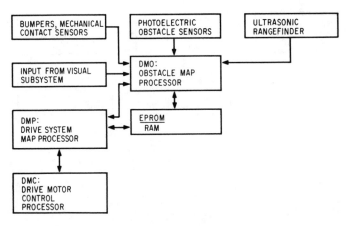

Fig. 20-1 *One way to organize the drive subsystem, including collision and obstacle avoidance and mapping. Input from the visual subsystem and other sensors help build a scrolling short-range obstacle map. This interacts with EPROM map data on a limited number of "known" or "home" rooms and map data on recently or currently learned "new" rooms. Data includes X-Y coordinates of doors, obstacles, power outlets, stairways, etc. Route selection, depending on the goal selected by the overseer processor, occurs here, and appropriate navigational data formatted for the drive motor power interfaces.*

light should be turned on, and the knowledge that the door is closed means the light should not be turned on. The simple circuit that accomplishes this switching task is a permanent manifestation of the refrigerator's ability to understand that knowing the door is open means turn on the light.

We can use microprocessors in a number of ways related to knowledge. In one sense, we can use them simply to gather knowledge without analyzing it, which is what might happen if we assigned a microprocessor to collect servo position information, for example, then pass it along as requested. In another sense, they can convert knowledge without understanding it—and if you believe that, you haven't been paying attention. In order to achieve any meaningful action with or upon data, the microprocessor (in concert with its programming) must understand what the data coming in means, at least in terms of what response should result from each stimulus. One excellent example of data conversion (knowledge conversion) is the conversion of mapping system data to a head's-up map.

There are several forms of manipulations we can perform, each designed to either increase the android's knowledge, convert the knowledge to a form with greater meaning, or increase the android's understanding of the meaning of the knowledge.

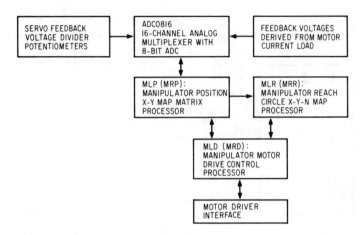

Fig. 20-2 *One way to organize the manipulator positioning subsystem. Servo and motor load data are digitized, devectored, and converted to an X-Y coordinate for each axis, an X-Y-Z coordinate overall. A parallel processor compares current servo positions with an end-of-travel lookup table and prepares reach area maps for each axis. The motor drive control selects the best way to reach the target and provides the appropriate motor drive signals. This is duplicated for each manipulator. Nomenclature for the left arm is shown in parentheses.*

And that's the key to organization for our processors. Our chain of command will be a chain of increasing sophistication of the knowledge being processed.

In many cases, as knowledge and understanding and meaning become increasingly sophisticated along one line, the android will need to coordinate with knowledge or meaning or understanding from another area, and a one- or two-way exchange of information across the channels will need to happen. For example, the visual system and the collision-avoidance system must both be consulted by the mapping system for it to gain knowledge of the android's environment.

Our problem, then, is to define the tasks facing the various android subsystems explicitly, to likewise define the division of these tasks between and among the several subsystems, and to detail the specific knowledge *and action* that is the responsibility of each subsystem to provide and perform—and, by the way, where it must go to pick up whatever knowledge it needs to do the job it's been assigned.

Because of the complexity potentially involved in decision making, the concept of dedicating whole processors to the role of decision maker is well worth considering. We can salt the structure with decision-maker microprocessors at any number of key points and use these processors as junior executives to coordinate the activities of groups of processors beneath them. As the android becomes more and more complex, these junior executives may have to report to senior executives. Ultimately, one processor has the final say—the overseer.

There are two fundamental approaches to organizing the task-assignment process. One is to work from the top down, deciding just what we want the android to be capable of, then determining how to accomplish our goals, step by step. The other is to work from the bottom up, starting with what sensors and effectors are available to the android and seeing what we can accomplish with them as we use more and more sophisticated techniques in analyzing, coordinating, or controlling them. In practice, we're likely to use both approaches, because as humans we have difficulty in *not* using every bit of knowledge available to us.

Shakespeare's plays often demonstrated the conflicts between the Apollonian and Dionysiac influences in our lives—our tendency to, on the one hand, be thinkers and seek knowledge and, on the other, to act. This same classic conflict will be a constant nemesis to our programming efforts and program execution.

We could build a hyperactive android, one with a strong tendency to keep in motion using only a minimal amount of understanding of its environment—just enough to stay out of trouble—and a minimum of curiosity. We also could build an overly pensive android, one with an urge to continue observing and investigating, but not very interested in doing anything. Actually, we'll want to build in a tendency to investigate before doing much of anything, then to do enough to bring up something new to investigate unless some specific action is required

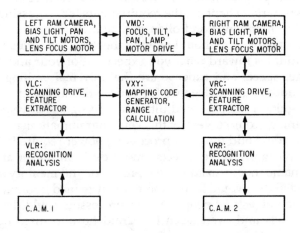

Fig. 20-3 *One way to approach task organization of the visual subsystem.*

because of a commanded or programmed priority. This procedure makes learning a priority. With proper heuristic programming that lets experience add to learning, we can add rehearsal to this list of priorities. As Davy Crockett suggested, the android can be sure what's right before going ahead.

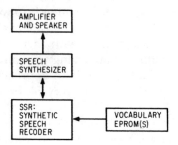

Fig. 20-4 *One way to organize the voicebox subsystem. Recoder receives data on word or phrase to be spoken, retrieves starting address and length of data block from "glossary" section of vocabulary EPROM, and drives speech synthesizer. This scheme functions for both fixed-vocabulary (word based) and universal (phoneme based) artificial speech.*

What we've discussed so far are the barest bones on which you will have the challenging task of building layers of knowledge, meaning, and understanding through the wise use of microprocessors, related hardware, and programming. As we discussed each of the android's subsystems in earlier chapters, the requirements for processing and control have been mentioned and often outlined. And that's just about as far as we can take you without knowing the specifics of your hardware and your attitudes toward what you expect out of your android.

Time, space, and expense are the only real limiting factors in developing additional intelligence for your android. For example, your speech recognition system could ultimately recognize any speaker and extract meaning from context with even ungrammatical users of language if you could add the extra processing power required. The visual system can recognize more objects, and a decision maker at this high level could name them. An additional processor and memory subsystem could be added that would bank experientially gained information about the recognized objects or people. Other processors could develop conversational skills and even teach the droid how to disco dance. The decisions are up to you.

And one of your first series of decisions is going to be which microprocessor to use for any given task. Too much capability could

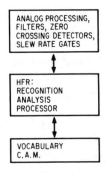

Fig. 20-5 *One way to organize the speech recognition subsystem. Comparatively narrow bandwidth filters reduce analog data to a few channels. Zero crossings in each channel are analyzed and coded by ranges. High slew wavefronts, as in plosives, are separately flagged. Processor compares input with vocabulary data in content addressable recognition memory. Recognized words are sent downstream to the overseer processor.*

cost money or unnecessary support hardware. The price for too little capability is frustration and the uneconomical option of patchwork make-do. Take careful note of the capabilities of the various microprocessors available, and choose wisely.

To help you decide, here are some sources for information for a number of microprocessors. The listing is not intended to be comprehensive. We assume you're already at least rudimentarily familiar with the most popular microprocessors—the 8080, the 6800, and the 6502. Here are some of their kith, kin, cousins, and offspring.

The 8048 from Intel (3065 Bowers Avenue, Santa Clara, CA 95051) is the flagchip of a line descended from the 8080. The 8048 includes on-chip ROM program memory and RAM. The 8748 substitutes EPROM for ROM. Still another version, the 8035, includes no on-chip program memory. These stout fellows all feature single 5-volt supply, an 8-bit CPU, 64 bytes of RAM, 27 I/O lines, an interval timer/event counter, an oscillator and clock driver, interrupt circuitry, a reset circuit, 8-level stacks, single-step operation, and two working register banks. The clock oscillator can be RC, crystal, or external. All instructions can be accomplished in either one or two cycles, with 2.5- and 5.0-microsecond versions available. Best of all is the repertoire of over 90 instructions. The 8048, 8748, and 8035 are all packaged in 40-pin DIPs.

The 8021 is a smaller, 28-pin microprocessor that uses a subset of the 8048 instruction set but offers increased I/O versatility. The 8022 adds interrupts, additional I/O, and linear A/D circuitry to the 8021. Write Intel for Application Note AP-56, *Designing with Intel's 8022 Microcomputer,* which includes a very interesting dc motor control application idea.

The 8049 is pin-compatible with the 8048 (etc.) but offers a number of advantages. On-chip ROM is expanded to 2 kilobytes; RAM to 128 bytes. It has facilities for extensive bit handling, and BCD as well as binary arithmetic. *Serial I/O and Math Utilities for the 8049 Microcomputer* is the title of Intel Application Note AP-49, available on request. The 8039 is an equivalent for the 8049 designed for use with off-chip program memory.

Adroit Electronics, Inc. (5 East Long Street, Suite 1012, Columbus, OH 43215) offers an 8035-based single board microcomputer with an 8-bit input port, an 8-bit output port, an 8-bit bidirectional port, handshake lines, crystal clock, provisions for 4K×8 program memory plus 1K×8 RAM. At this writing, suggested U.S. resale is a surprisingly low $99.95.

Intel also offers the 8041 (1K×8 ROM, 64×8 RAM, 18 programmable I/O pins) and 8741 (same, except ROM area is EPROM) single-chip microcomputers with facilities designed to make it especially useful as an interface to peripherals. In other words, it is designed for very effective, very efficient use as a slave processor in a multiprocessing system—especially significant in light of our android's requirements.

Intel also offers a number of support circuit ICs, including memory and I/O expanders and much more. This family of microcomputer products, MCS-48™, *could* be used exclusively to provide all necessary processor functions for an android, if desired

There are pin-compatible versions of the 8748 and 8741 available in CMOS, which both reduces the power dissipation and extends the operating temperature range. These devices, the 87C48 and 87C41 (officially, IM87C48IJL and IM87C41IJL) are available from Intersil, Inc. (10710 North Tantau Avenue, Cupertino, CA 95014).

The 8048 also is available from Signetics (a subsidiary of U.S. Philips Corporation, P.O. Box 9052, 811 East Arques Avenue, Sunnyvale, CA 94086).

National Semiconductor (2900 Semiconductor Drive, Santa Clara, CA 95051) offers a wide "48 Series" selection of single chip microcomputers. The INS8048, INS8039, and INS8049 are equivalent to the 8048, 8039, and 8049, respectively. The INS8050 is similar, but offers 4K of ROM and ¼K of RAM; the INS8040 is a ROMless version. Also, National's INS8036 is a ROMless INS8048.

National also offers the INS8072, which includes built-in multiprocessing logic (!), bus control logic and DMA logic, 16-bit arithmetic instructions with its 8-bit CPU, plus search and compare functions. Sixty-four bytes of RAM and 2½K bytes of ROM are on-chip—the INS8070 is a ROMless equivalent.

The newest microprocessor in National's line at this writing is the NSC800 family of P²CMOS processors and support circuits.

(P²CMOS is a proprietary process that yields CMOS circuits that operate in NMOS speeds.) The intriguing feature of the NSC800 is that it offers both the architecture of the Intel 8085 and the very powerful instruction set of the Zilog Z-80.

In addition, a number of other features have been designed in, not the least of which are a direct result of the CMOS technology. Power consumption is about 20-to-1 less than an NMOS processor. It uses a single 3- to 12-VDC supply. It offers a clock generator, RAM refresh circuitry, five interrupt request lines, a system controller, a power-saving standby circuit, and a reset circuit offering both a Schmitt trigger input and a reset output available to the rest of a system.

National has a full line of peripheral and support circuitry available for the NSC800. Unfortunately, the three-chip basis for a system carries an initial 100+ quantity price of about $175. As with all things electronic, we can hope for this to come down as volume goes up— meaning the volume National manufactures.

Speaking of the Z-80, those applications that need its gangbuster instruction set can't do better. This little gem from Zilog (10460 Bubb Road, Cupertino, CA 95014) has been offered for sale by parts stores for under $15. In addition to the 78 instructions the 8080A uses, the Z-80 offers 80 more. These include 4-bit, 8-bit, and 16-bit operations; indexed, relative, and bit address modes; 17 internal registers; three interrupt response modes plus a nonmaskable interrupt; dynamic RAM support; and single-supply 5-VDC operation with TTL compatibility at all pins.

Zilog also makes the Z-8 (which is now second-sourced by Synertek), a single-chip microcomputer offspring of the Z-80.

One considerable advantage of machine language programming with the Z-80 (as compared to other CPUs) is that a great deal of it can be developed on a Radio Shack TRS-80. There is a great deal of machine language programming for the TRS-80 available and more being developed all the time. Inevitably, much of this will be applicable to android subsystems.

Available software is a *major* advantage in saving development time. And I'm told that no system in history has had more software developed for it than the PDP-8. (PDP-8 is a registered trademark of Digital Equipment Corporation, Maynard, MA.) There's a CMOS 12-bit microprocessor that's software compatible with the PDP-8. It's the IM6100 from Intersil, which is second sourced as the HM-6100 by Harris Semiconductor (A Division of Harris Corporation, 2016 Quail Street, Newport Beach, CA 92660). Intersil makes available the 6950 Intercept Jr. tutorial single board computer based on the 6100; Harris offers the HB-61000 Micro-12 single board 6100-based processor, both complete with keyboard and display. In addition, Cybertek, Inc. (P.O. Box 3467,

Seminole, FL 33542) offers LP-12, a 6100-based series of small boards that combine into a flexible processing system primarily intended for battery-operated applications.

As long as we're on CMOS, the most mature 8-bit CMOS microprocessors I know of are the COSMAC 1802 series, available at the component, board, and system level from RCA Corporation (RCA Solid State, Box 3200, Somerville, NJ 08876); it is second sourced by Hughes and others. The 1802 also is the basis for the Elf and Super Elf, which are available from Quest Electronics (P.O. Box 4430C, Santa Clara, CA 95054) and Netronics R&D Ltd. (333 Litchfield Road, New Milford, CT 06776).

There are many more sources for microcomputers and boards than we can hope to cover here. And, frankly, the choice of a specific processor is not as critical a selection as most of the manufacturers would have you believe. Since each processor will be dedicated to the tasks you assign it, once you know it can accomplish those tasks, any additional, unused capabilities are meaningless. It could well be that the most powerful microprocessors also are the least expensive, easiest to work with, and so on, and that's fine. My point is that making your choice is not as difficult as you might think.

Before we leave the subject, there are a number of excellent articles that can contribute to our understanding and appreciation of many of the subtleties of the design tasks facing us.

Joseph P. Barthmaier of Intel Corporation, Hillsboro, Oregon, authored an interesting article for the February 1980 *Computer Design* entitled "Multiprocessing System Mixes 8- and 16-bit Microcomputers." This article provides an excellent review of the advantages of multiprocessing, like resource sharing between processors; a few of the techniques, like interprocessor interrupts, necessary to accomplish fast communication; hardware discussions; and more.

The June 1977 "Bionics Special" issue of *Interface Age* provided both intriguing articles and excellent references for further investigation of computer control of prosthetic devices. Amos Freedy, John Lyman, Moshe Solomonow of the Biotechnical Laboratory of the University of California at Los Angeles authored "A Microcomputer-Aided Prosthesis Control System," which suggests control techniques adaptable to the android's arms and hands. Daniel Graupe, Ph.D., then the Russell S. Springer Visiting Professor at the Department of Mechanical Engineering at the University of California at Berkeley, suggests another control approach in his article, "Microprocessor Controls Prosthesis via EMG Signal."

Here are a few of the references cited by these gentlemen, chosen for their applicability to our needs:

- "Studies and Developments of Heuristic End Point Control for Artificial Upper Limbs" by Zadaca, Lyman, and Freedy. *University of California, Los Angeles, School of Engineering and Applied Science Report UCLA-ENG-74-79; Biotechnology Laboratory Technical Report Number 54, 1974.*
- "Adaptive Aiding in Artificial Arms Control" by Freedy, Hull, and Lyman. *Bulletin of Prosthetics Research, BPR 10-16:3-15, Fall, 1971.*
- *Optimization by Vector Space Methods* by D. G. Luenberger. New York: John Wiley and Sons, 1969.

By far, of course, the best works on android intelligence and the applications of microprocessors to android subsystems are those yet to be written. In this grayest of all areas of android design, I can only hope that my few notes have planted a few seeds in the right human minds to speed the day when these articles will appear.

21
Letters

It's a rare luxury for the author of a book to have received correspondence on his work before it appears. But thanks to an article that appeared in *Radio-Electronics* early in 1980 and an early item that appeared in *Electronic Engineering Times*, I've had a few letters already. They are condensed or extracted here, for the express purpose of making sure that the thoughts in this book are not purely my own.

The probability that later editions of this book will be expanded and updated suggests that we develop an open forum for thinking on the subject of android design. You are invited to forward your thoughts to the author in care of Hayden Book Company, Inc. (50 Essex Street, Rochelle Park, NJ 07662).

Mr. Joseph Coppola of Canastota, New York, writes:

I have been designing and building robots as a hobby for twenty-one years . . . (since) back in the days when people thought you were crazy if you told them you built robots.

There are a number of flaws in Mr. Weinstein's approach to his mechanical butler. For one thing, tracked vehicles of any weight over a few pounds tend to be very destructive to carpets, throw rugs, almost any kind of household flooring. My family was ready to disown me after my last tracked machine, and it only weighed fifty-seven pounds. The best solution I have found for heavy machines in the house has been three or four independently driven wheels at least four and one-quarter inches in diameter, all of which turn at the same time for cornering. This allows the robot to rotate about its own vertical axis, eliminating the weak spot of the "Lazy Susan" joint and allowing the machine to turn in its own length. One can reduce the base of the machine to an eighteen inch square using this technique. When additional stability is required—as for lifting, etc.—hydraulic or motorized outriggers can extend this base to keep the machine's center of gravity contained within its footprint.

I enjoyed the article and would like to see more of the same. Thank you for your time.

Mr. Coppola's points about track are well taken. The Uniroyal belting mentioned in the text is less guilty of damage to flooring than many other kinds of belting and most other track materials, but there is still some chance of wear and tear, scuffing and black "heel" marks. The planetary triangular wheel drive should eliminate most of these problems.

The turntable bearings used in the android are very substantial and should not prove a "weak point" of the mechanical design; however, the largest part of the android's weight is located below these bearings and so are not borne by them.

And, of course, the idea of outriggers is intriguing.

Professor E. Kafrissen of the Department of Computer Science, Electrical Engineering, and Technology at the New York Institute of Technology doesn't quite believe my description of power requirements:

> It may be of interest that we are in the process of building an android whose base is virtually identical to (your) design.
> I would like to make a few comments that might be relevant:
> a. Whether or not the android can climb steps with tracks is as much a traction problem as anything else. Military tanks are required to climb a 60% slope (33°). They certainly have the power to climb 45°, but I doubt they could obtain sufficient traction. To check this, I tried an experiment with a 1/15 scale model tank and found it could not obtain traction in a slope greater than 30°.
> b. Your figures on power required to move the android should be rechecked. Unfortunately, I do not have a copy of (your figures) in front of me, but I believe—or hope—that an android could be moved with less power than you say. For example, a 4 HP motorcylce can move a 250 pound person at better than 40 mph. I had such a motorcycle, so I know it is true. Hence, 2 HP should be enough to move a 400 pound android at 7 or 8 mph. However, if you accept your HP figures, then the battery drain would be much higher that you stated.
> A 12 Volt 40 Amp-hour battery would be drained in about 10 minutes if it had to deliver 2 HP.

We saw in our own calculations that the major contributing factors to drive motor horsepower requirements weren't so much top speed as acceleration and steepness of climb. If not for the stair-climbing need, we wouldn't need anywhere near the power. Fortunately, we don't need to provide this much power continuously, so the picture for our batteries is not quite so bleak.

The next letter was an interesting piece of "hate" mail sent to *Radio-Electronics* from H.O. Olson:

If Martin Bradley Weinstein does not have any more to offer than he did in this article, keep him on something simple, like how to build a good logic monitor.

If you have any more articles on androids, start at the beginning with robots. Start with the platform, the wheels, the best motors to start with, have the writer build it, and then put down on paper exactly what he did and used to do it.

Industry uses many different kinds of robots—add this information to your article.

There are electric trains that move parts through factories by following wires in the floor. So have the robot be able to follow a predetermined route through a house.

You could use low voltage wiring and then you could have each item in the house set up with its own frequency on each door or other item.

(There are) remote controls for lights and appliances. Have the robot use them to open and close doors.

There are door latches that are operated by low voltage, so you could use them with the robot. He could also turn on and off the lights and other appliances and motors needed to move the doors on tracks.

Then set up something simple for him (the robot) to do, like moving around in your house when you are away from home to keep the thieves away.

Then, when your robot starts to think for himself (built-in timers or computer) you could . . . give it voice-activated speech (in English only), and there is no end to what you could come up with.

Better luck in your future articles.

Gosh, I hope so.

At the other end of the correspondence spectrum is this thought-provoking letter from Jim McBeath of Redondo Beach, California:

Here are some thoughts about my approach towards creating an android.

My goal is to create an android that is superficially indistinguishable from a human. By this I mean that I could introduce him (or her) to a person who does not know that he is talking to an android, and that it would take at least a handful of seconds of interaction before he realized that. Just about everyone that I have told this to has assured me that I'm crazy, or at least has made it clear that he thinks it is not possible. My primary contention is that technology is advancing at an ever-increasing rate, so that within my lifetime I will see things at least as amazing to us as a pocket computer would be to someone from the turn of the century. Since I am at present a young man, I hope to see at least another 50 years before I go. In that time I would not be overly surprised to see the following occur:

- A unified gravity/E&M field theory, which might lead to the invention of antigravity devices and force fields.
- Room temperature superconductors, or at a minimum, superconductors in the temperature region of liquid nitrogen.
- Fusion power. Possibly a fusion or fission power source which can be housed in a very small container (one to ten liters).
- Advances in material and chemical engineering, bringing out new substances such as plastics and Teflon, and very high strength cables (made of bundled monofilaments). Also, a lot of heretofore-unknown synthetics.

I could probably continue the above list, but I think you can see the point. Imagine going back a hundred years and trying to figure out how you would send a man to the moon. It would be quite a task—most of the things required were not even close to being on the drawing board yet. And at the rate of increase of technology, it does not seem unreasonable that technology 50 years from now will be at least as far past us as we are from 100 years ago.

By now I may have convinced you that I'm crazy, or I may have convinced you that my ideas are feasible; however, whether or not it really is possible, the important thing is that I think it is—so let us get on with how I plan on going about this nontrivial task.

As indicated above, I don't really plan on starting this thing for another forty or fifty years. I'm not even sure whether that is sufficient time to learn what I will need to know, but I can certainly try. The area which strikes me as being the most difficult is the understanding of human communication so that people can communicate with my android. In the meantime, while waiting for all this fantastic technology to come along, I can try to work on some of the subsystems. The problems I expect to run into are going to be theory-related problems, not technology limitations. As an example, take seeing. I think that the hardware will exist to do as much processing as I want, but the algorithms for object and motion recognition may not be sufficiently developed. I'm not saying the hardware will necessarily be as sophisticated as the human eye and visual system,

I'm just saying that the weak link in that part of the system will be the algorithms to process the data.

As mentioned above, the biggest problem that I see is the communication problem. This includes nonverbal communication and intonation as well as words. A subsystem which could possibly be built earlier would be a disembodied voice. I'd like a "secretary" who could answer my phone while I'm away, take messages, give messages to the appropriate people and perhaps even chat a little. This subsystem would contain basically all the "intelligence" that would have to go into the walking version. By this I mean all the problems of artificial intelligence. The functions that the body has to perform that the voice would not are not intelligence functions. There may be some learning processes, but basically there would not be any real intelligence.

The communication problem as I described it above included the aspect of intelligence. This is the one area where I think the hardware might possibly not exist at the complexity/density I want. There is still the tremendous job of programming, but my approach would be as much as possible to create a learning machine so as to need as little initial programming as possible. In particular, one of the critical aspects is that the brain be able to learn how to learn. A high degree of homogeneity would probably be desired in order to make things simpler. Thus, initially, only a small portion of the brain hardware would be used, which would grow larger as the android learned more complicated methods of processing.

Another major problem would be locomotion. I would try to design a body with internal joints very similar to a human; i.e, a ball and socket at the hip and shoulder, simple hinges at the elbows and knees, etc. The joints would be controlled by a separate "mover" for each degree of freedom. By *mover* I mean a motor, magnet, or whatever is available at the time. Depending on the mover, there may or may not have to be the flexors and extensors as in the human. The movers would be controlled by a good deal of processing hardware. Probably each mover would have the equivalent of a medium to large microprocessor associated with it (large by today's standards, but small by tomorrow's). The movers would be gathered into local nodes, one at each joint. The local node would have another level of processing, which would be similar to a minicomputer of today. Then the local nodes would be collected into regional nodes, say one at each arm and leg, which would have yet another level of processing of power similar to a reasonably large computer of today. Then, finally, the regional nodes would feed into the "brain," which will be of considerably higher power than the largest of today's computers. Of course, the brain has a lot of other things to do besides move, so only a portion of it would be used to control motion. There would possibly be a reflex center between the brain and the regional nodes. There would be a great deal of feedback used—visual detection of motion might be used, as well as inertial sensors (equivalent to the inner ear in a human), which could be scattered through the body. This external feedback, as well as internal feedback (internal position indicators) would be used to determine the response of the movers. As you can see, I envision a lot of distributed parallel and pipeline processing. There's an awful lot of programming to think about.

The above descriptions are necessarily vague because I don't know what tools will be around to implement them. There are a lot of other human qualities that could be emulated, such as taste and smell, but I have not paid much attention to them because they are not as important for my goal of "superficially indistinguishable." As with other aspects, I would rely heavily on the use of distributed processing.

Communication among the distributed control nodes would depend on the data rate necessary. If a high data rate is needed, optic fibers could be used. If there is a low enough data transfer rate, much wire could be saved by using the net approach—all data is moved around on a single bus, with everyone looking at it. Preferably, all data would be on a single wire or fiber.

There is one more big problem, which is the logistics of how I am going to go about doing this, which I don't know either— but I figure something will come along, just like on the technical side of it (always the optimist!). I expect this to be a very expensive project, using state-of-the-art tools. I would estimate that the cost (in today's dollars) would be at least a few million dollars. This would have to be a research project (government?) with a team of people working on it. If the team were to form today, I would feel that my knowledge is not quite sufficient to qualify me (although I certainly would try!). In another 10 years or so, yes; however, I don't personally know anyone else that I think would qualify at present, although I'm sure there are some, and I am confident that I am at least as qualified as anyone else my age. And since I plan on continuing my education in that direction, I will be in an excellent position when the team finally does form (as I'm sure it will!).

Well, Marty, I hope this letter has given you some ideas, or at least something to chuckle about. I don't mind if you think

I'm crazy as long as you don't think I'm REALLY crazy. If it does turn out to be impossible, well, that's too bad—but none-theless it is very interesting to consider.

Well, Jim, everyone who sees the future has to be a little crazy—it's required. Most of us interested in android design often wonder why we are. And we think about things like—if androids aren't atheists, will *they* believe in *us*?

Your estimation of processing requirements is a few powers of ten out of whack and your knowledge of materials physics less than comprehensive, but these are just means in an inquiry where the ends are much more interesting.

Thank you at least for reminding us that the tin motor boxes we'll be building aren't the last step in a process, but a first step in an evolution. In a way, you know, that makes the task ahead all the more frightening, and all the more a challenge.

22
The Beginning

I am no pioneer. Some of the work behind this book is original, but most is borrowed from the many disciplines involved in designing as interdisciplinary a project as an android.

So my role is substantially that of a collector—of literature, of names and addresses, of scraps of information, of snatches of opinions and attitudes. In recording them, I have worked long and busy nights over the course of many, many months.

Chances are that you won't be satisfied with this book. It tries to explain or describe so many items of interest that it can't do a thorough job with any of them. That's why there are so many company names and addresses throughout. Write them for information. They can do a much more thorough job than I can because they'll keep answering as long as you keep asking. And this way, you're reading a volume that you can hold in a hand, not a series that eats up a shelf.

From the very first, this text has been intended to give you hints, ideas, and information that can help you design your own android. It is not a do-it-yourself book of plans, not a step-by-step construction guide, not so simple that a five-year-old can do it.

Now it's time for you to go to work. Start with pencil and paper, a drafting set, and a very open mind. Plan and plan again before you touch even one piece of hardware. Guaranteed, Murphy's Laws have never been more strictly enforced.

Design modularly. The ramera you build now to work with one planned configuration can be worth just so much scrap when you change to another. Assemble small, but make disassembly easy because you're going to have to make changes and repairs from the first to the last.

But chin up, Bunkie, because you, yes you, can really and truly build an android if you decide to. This is the generation that will see the birth of androids. My generation. Your generation. Their generation.

Welcome to our world, friend androids. Come and go in peace. May no one ever exploit you. You can do much to lessen our pain and

239

suffering. And that alone is enough to make the pain and suffering we must go through to bring you to life so very worthwhile.

This is the beginning.

And with those words, I leave you on your own.

Appendix

FOR FURTHER READING ON SPEECH SYNTHESIS

ARNOLD, WILLIAM F. "Signal Processor for Speech Synthesis Is Programmable," *Electronics*, October 11, 1979.

BRANTINGHAM, LARRY. "Speech Synthesis with Linear Predictive Coding," *Interface Age*, June 1979.

CIARCIA, STEVE. "Talk to Me!—Add a Voice to Your Computer for $35." *Byte*, June 1978.

ERMAN, LEE, ed. *IEEE Symposium on Speech Recognition*, Contributed Papers, IEEE Catalog #74CH0878–9AE, April 1974.

FLANAGAN, J.L. *Speech Analysis, Synthesis and Perception*, 2nd edition. New York: Springer–Verlag, 1972.

FLANAGAN, J.L., and RABINER, L.R. eds. *Speech Synthesis*. Benchmark Papers in Acoustics Series. New York: Academic Press, 1973.

LEHISTE, ILSE, ed. *Readings in Acoustic Phonetics*. Cambridge, Mass.: MIT Press, 1967.

MARKEL, J.D., and GRAY, A.H., JR. *Linear Prediction of Speech*. New York: Springer–Verlag, 1976.

MOSCHYTZ, GEORGE S. *Linear Integrated Networks Design*. New York: Van Nostrand, 1975.

NORTH, STEVE. "Mountain Hardware SuperTalker," *Creative Computing*, October 1979.

RICE, D. LLOYD. "Friends, Humans, and Countryrobots: Lend Me Your Ears," *Byte*, August 1976.

SHERWOOD, BRUCE A. "The Computer Speaks—Rapid Speech Synthesis from Printed Text Input Could Accommodate an Unlimited Vocabulary," *IEEE Spectrum*, August 1979.

WEINSTEIN, MARTIN BRADLEY. "Machines that *Can* Talk," *Radio-Electronics*, March and April, 1980.

WIGGINS, RICHARD, and BRANTINGHAM, LARRY. "Three-Chip System Synthesizes Human Speech," *Electronics*, August 31, 1978.

241

FOR FURTHER READING ON SPEECH RECOGNITION

BEHRENS, CHARLES W. " 'Speak Master, and I Will Obey'—Part II: The Possibility of Appliances that Listen and Respond," *Appliance Manufacturer*, December 1978.

BEZDEL and BRIDLE. "Speech Recognition Using Zero Crossing Measurements and Sequence Information," *Proceedings of the IEEE*, Vol. 116, 1969.

BRODDIE, JAMES R. "Speech Recognition for a Personal Computer System," *Byte*.

COX, R.B., and MARTIN, T.B. "Speak and the Machines Obey," *Industrial Research*, November 15,1975.

DAVIS, BIDDULPH, and BALASHEK. "Automatic Recognition of Spoken Digits," *Journal of the Acoustical Society of America*, November 1952.

ENEA, HORACE, and REYKJALIN, JOHN. "Introducing Speechlab, the First Hobbyist Vocal Interface for a Computer!" *Popular Electronics*, May 1977.

GEORGIOU, BILL. "Give an Ear to Your Computer," *Byte*, June 1978.

GILLI and MEO. "Sequential System for Recognizing Spoken Digits in Real Time," *Acustica*, Vol. 19, 1967.

GRUNZA, E.F. "Electronic Speech Recognition," *Medical Electronics*, February 1978.

ITAKURA, F. "Minimum Prediction Residual Principle Applied to Speech Recognition," *IEEE Transactions on Acoustics, Speech and Signal Processing*, Vol. 23, 1975.

LAMB, SYDNEY M., and VANDERSLICE, RALPH. "Recognition Memory: Low-Cost Content Addressable Parallel Processor for Speech Data Manipulation," paper presented at the Acoustical Society of America and Acoustical Society of Japan Joint Meeting Session, *Speech Communication Meets the IC Revolution*, November 1978.

LINDGREN, NILO. "Machine Recognition of Human Language," *IEEE Spectrum*, March and April, 1965.

MARTIN, T.B. "One Way to Talk to Computers," *IEEE Spectrum*, May 1977.

———. "Practical Applications of Voice Input to Machines," *Proceedings of the IEEE*, April 1976.

OTTEN and KLAUS. "Approaches to the Machine Recognition of Conversational Speech," in *Advances in Computers*. New York: Academic Press, 1971.

PLUTE, M. F., and COX, R. B. "Voice Programming—A New Dimension to NC," *Proceedings of the Technical Conference, MC/CAM Expo '75*, May 1975.

PURTON. "Speech Recognition Using Autocorrelation Analysis," *IEEE Transactions on Audio and Electroacoustics*, June 1968.

RABINER, L.R., and SCHAFER, R.W. "Digital Techniques for Computer Voice Response: Implementations and Applications," *Proceedings of the IEEE*, Vol. 64, 1976.

ROSENBERG, A.E. and ITAKURA, F. "Evaluation of an Automatic Word Recognition System over Dialed-Up Telephone Lines," *Journal of the Acoustic Society of America*, Vol. 60, supplement 1, 1976.

ROSENBERG, A.E., and SCHMIDT, C.E. "Automatic Recognition of Spoken Spelled Names for Obtaining Directory Listings," *The Bell System Technical Journal*, October 1979.

ROSENTHAL, L.H., et al. "A Multiline Computer Voice Response System Using ADPCM Coded Speech," *IEEE Transactions on Acoustics, Speech and Signal Processing*, Vol. 22, 1974.

ROSS. "A Limited-Vocabulary Adaptive Speech Recognition System," *Journal of the Audio Engineering Society*, October 1967.

SIMMONS, E. JOSEPH, JR. "Speech Recognition Technology," *Computer Design*, June 1979.

STEWART, J.L. *The Bionic Ear*. Santa Maria, Calif.: Covox, 1979.

TEACHER, KELLETT, and FOCHT. "Experimental, Limited Vocabulary Speech Recognizer," *IEEE Transactions on Audio and Electroacoustics*, September 1967.

THORDARSON, P. "Design Guidelines for a Computer Voice Response System," *Computer Design*, November 1977.

WHITE and GEORGE. "Speech Recognition: A Tutorial Overview," *IEEE Computer*, May 1976.

WOOD, N. "Need Those NC Tapes Quick and Easy? Talk to Your Computer," *Machine and Tool Blue Book*, May 1976.

Index

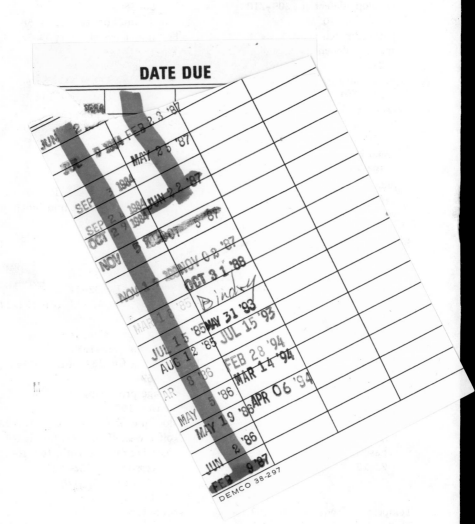